Plates

K. Bhaskar · T. K. Varadan

Plates

Theories and Applications

Ane Books
Pvt. Ltd.

K. Bhaskar
Department of Aerospace Engineering
Indian Institute of Technology, Madras,
Chennai, India

T. K. Varadan
Department of Aerospace Engineering
Indian Institute of Technology, Madras,
Chennai, India

ISBN 978-3-030-69426-5   ISBN 978-3-030-69424-1   (eBook)
https://doi.org/10.1007/978-3-030-69424-1

Jointly published with ANE Books Pvt. Ltd
In addition to this printed edition, there is a local printed edition of this work available via Ane Books in
South Asia (India, Pakistan, Sri Lanka, Bangladesh, Nepal and Bhutan) and Africa (all countries in the
African subcontinent).
ISBN of the Co-Publisher's edition: 978-9-382-12702-4

This Springer imprint is published by the registered company Springer Nature Switzerland AG
The registered company address is: Gewerbestrasse 11, 6330 Cham, Switzerland

Preface

Plates are encountered in several forms—starting from the simple uniform, thin, homogeneous metallic structure to more efficient and durable alternatives involving features such as variable thickness, lamination, sandwich construction, fibre reinforcement, functional gradation, and moderately-thick to very-thick geometry. Correspondingly, several theoretical models are employed for their analysis and design starting from the classical thin plate theory to alternatives obtained by slight modifications to incorporate the appropriate complicating effects, or by putting forth fundamentally different initial assumptions. This book is an attempt to capture the essentials of this development and to present it such that the reader can obtain a quick understanding and overview of the subject area of plate structures.

For the sake of convenience, the book is divided into two parts. The first part is devoted to a complete, self-sufficient exposition of the classical theory with several solved and unsolved problems to demonstrate the various steps involved in the analysis of static flexure or free vibrations or stability. This introductory part is expected to be useful as a text-book for post-graduate engineering students and research scholars enrolled in a first-level course on plates.

The second part of the book is devoted to advanced theories and applications, with each complicating feature discussed separately and with appropriate literature cited for further study. This presentation is at an advanced level as compared to the first part and is expected to be useful for research scholars as well as practising engineers.

Chennai, India

K. Bhaskar
T. K. Varadan

Contents

About the Authors

Prof. K. Bhaskar has been with Indian Institute of Technology, Madras, since 1992. As part of the Department of Aerospace Engineering, he teaches courses related to Solid Mechanics and Elasticity. His research contributions are primarily related to theoretical modeling of thick laminated structures, with around fifty publications in refereed international journals.

Prof. T. K. Varadan was with the Department of Aerospace Engineering, Indian Institute of Technology, Madras, for more than 35 years before retiring in 2001. Besides teaching a wide variety of courses related to Structural Mechanics and Aircraft Design, he has made significant research contributions in the areas of Non-linear Vibrations and Composite Structures, with more than one hundred publications in refereed international and national journals.

Part I
Classical Theory and Straightforward Applications

Introduction to Part I

Plate-like structures are encountered in many fields of engineering. Examples include wing panels and rocket fins in aerospace engineering, building floors and walls in civil engineering, automotive body panels and disk wheels in mechanical engineering, ship decks and off-shore platforms in ocean engineering, and printed circuit boards in electronics. It is necessary to understand the behaviour of these plate structures— their deflections and stresses due to various static and dynamic loads, their tendency to resonate, and their susceptibility to buckling—so that they can be designed to withstand the service conditions. Such an analysis should be fairly accurate and also simple. Hence, an engineering theory, satisfying these seemingly contradicting requirements, is very important. The simplest theory developed for this purpose, known as the *Thin Plate Theory* or the *Classical Plate Theory*, is the topic of the first part of this book, which includes illustrative applications involving homogeneous isotropic plates of uniform thickness.

For the sake of better comprehension, several conceptual questions and unsolved problems are given in each chapter; a corresponding solution manual giving hints and final answers is included as an Appendix.

Chapter 1
Definition of a Thin Plate

As a prelude and to place the engineering theory to be developed later in proper perspective, a simple test problem is analysed in this chapter using the rigorous approach of the theory of elasticity, i.e. by considering the plate as a three-dimensional solid. The assumptions that lead to thin plate theory are brought out as natural inferences from the results of this three-dimensional analysis. A brief outline of the field variables and governing equations of elasticity is also included.

1.1 The Elasticity Approach

The main difference between the elasticity approach and conventional engineering analysis based on the mechanics-of-materials approach is the inclusion of a suitable hypothesis regarding the geometry of deformation in the latter. For instance, the engineering theory of beams is based on Euler–Bernoulli hypothesis regarding the preservation of plane cross sections during deformation; the corresponding analysis using the theory of elasticity does not require this hypothesis.

The theory of elasticity is based on the use of the concepts of equilibrium, continuum and a material constitutive relationship to analyse a structure, i.e. to obtain the displacements, strains and stresses at any point within it when the loads and support conditions are specified. The theory can be stated in a mathematical form in terms of the field equations, viz. the equations of equilibrium, the strain–displacement relations and the stress–strain law; these equations are given here with reference to Cartesian *x-y-z* axes for a *linearly elastic isotropic body undergoing small deformations in the absence of body forces*. The equations are in terms of

© The Author(s) 2021
K. Bhaskar and T. K. Varadan, *Plates*,
https://doi.org/10.1007/978-3-030-69424-1_1

Fig. 1.1 Positive stress
components

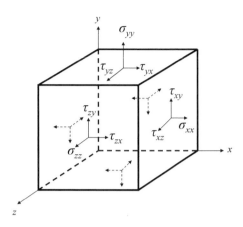

the field variables, viz. the three displacements u, v and w along x, y and z directions, respectively, the nine components of the strain tensor and the nine components of the stress tensor. Normal stresses and strains are usually denoted by σ_{ii} and
ε_{ii}, respectively, while shear stresses and strains are denoted by τ_{ij} and γ_{ij}, with
$i, j = x, y, z$. The first subscript in all these quantities denotes the direction of the
normal to the plane under consideration, and the second subscript denotes the direction of the stress or strain. If the outward normal to a surface is in a positive (negative)
coordinate direction, the corresponding stresses are taken as positive when acting in
the positive (negative) coordinate directions. This is the sign convention commonly
adopted and is as shown in Fig. 1.1; a corresponding sign convention holds good for
strains.

  It should be noted that only six stress (and six strain) components are independent
since $\tau_{ij=}\tau_{ji}$ (and $\gamma_{ij=}\gamma_{ji}$) by the principle of complementary shear. In this book, the
normal stresses and strains are denoted using single subscripts, i.e. σ_i and ε_i instead
of σ_{ii} and ε_{ii}.

  The field equations are

Equilibrium Equations

$$\sigma_{x,x} + \tau_{xy,y} + \tau_{xz,z} = 0$$
$$\tau_{xy,x} + \sigma_{y,y} + \tau_{yz,z} = 0$$
$$\tau_{xz,x} + \tau_{yz,y} + \sigma_{z,z} = 0 \tag{1.1}$$

Strain–Displacement Relations

$$\varepsilon_x = u_{,x} \quad \gamma_{yz} = v_{,z} + w_{,y}$$
$$\varepsilon_y = v_{,y} \quad \gamma_{xz} = u_{,z} + w_{,x} \tag{1.2}$$
$$\varepsilon_z = w_{,z} \quad \gamma_{xy} = u_{,y} + v_{,x}$$

Stress–Strain Law

$$\sigma_i = 2G\varepsilon_i + \lambda e, \quad \tau_{ij} = G\gamma_{ij} \quad \text{with } i, j = x, y, z \tag{1.3}$$

These field equations are supplemented by the boundary conditions, i.e. the mathematical description of the supports and the applied loads.

In the above equations, G and λ are Lamè's constants, e is the volumetric strain given by $\varepsilon_x + \varepsilon_y + \varepsilon_z$, and a subscript comma is used to denote differentiation. In terms of the Young's modulus E and Poisson's ratio μ, Lamè's constants are given by

$$\lambda = \frac{\mu E}{(1 + \mu)(1 - 2\mu)} \quad G = \frac{E}{2(1 + \mu)} \tag{1.4}$$

The solution of the elasticity problem can be carried out by using the *displacement approach*, wherein the displacements u, v and w are first solved for, and then the strains and stresses are evaluated using Eqs. (1.2) and (1.3), or the *stress approach*, wherein the stresses are first obtained and the strains and displacements subsequently, or a *mixed approach*.

The field equations given above are sufficient for the displacement approach, while, in addition, certain conditions are required in the other approaches to ensure the continuum nature of the structure after deformation; these *compatibility* conditions will be introduced later in the book as and when a need for them arises.

1.2 A Test Problem

One of the problems amenable to exact analysis using the elasticity approach is considered here. An infinitely long prismatic body of rectangular cross section as shown in Fig. 1.2 is supported at the two longitudinal edges such that

$$w = \sigma_x = 0 \text{ at } x = 0 \text{ and } a, \text{ for all } y \text{ and } z \tag{1.5}$$

Fig. 1.2 Test problem

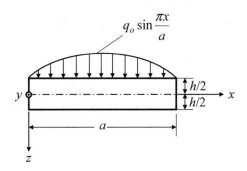

This boundary condition, which we shall encounter later on also, corresponds to supporting the end planes ($x = 0$ and $x = a$) by *shear diaphragms* attached to them—a shear diaphragm being defined as one that completely restrains displacements in its own plane, but fully permits the out-of-plane displacement. Obviously, this condition corresponds to a simple support in the limiting case of a thin plate.

The body is subjected to sinusoidal transverse loading applied on the top surface (Fig. 1.2). If this load is assumed to be independent of y, then the body undergoes cylindrical bending such that the displacements, strains and stresses do not vary along y, and the problem reduces to one of the cross-sectional plane. It is further assumed that the ends of the infinitely long strip are restrained in the axial direction so that $v = 0$ everywhere, reducing the problem to one of plane strain.

Corresponding to the lateral surfaces, the boundary conditions are

$$\sigma_z = -q_o \sin \frac{\pi x}{a} \quad \text{and} \quad \tau_{xz} = 0 \text{ at}$$
$$z = -h/2, \quad \text{for all } x \text{ and } y$$
$$\sigma_z = \tau_{xz} = 0 \quad \text{at } z = h/2, \quad \text{for all } x \text{ and } y \tag{1.6}$$

The other equations to be satisfied are the field Eqs. (1.1)–(1.3) simplified for the case of cylindrical bending as given below.

$$\sigma_{x,x} + \tau_{xz,z} = 0$$
$$\tau_{xz,x} + \sigma_{z,z} = 0 \tag{1.7}$$

$$\varepsilon_x = u_{,x}$$
$$\varepsilon_z = w_{,z}$$
$$\gamma_{xz} = u_{,z} + w_{,x} \tag{1.8}$$

$$\sigma_x = 2G\varepsilon_x + \lambda(\varepsilon_x + \varepsilon_z)$$
$$\sigma_z = 2G\varepsilon_z + \lambda(\varepsilon_x + \varepsilon_z)$$
$$\tau_{xz} = G\gamma_{xz} \tag{1.9}$$

Following the displacement approach, the governing equations are obtained by the use of Eqs. (1.8) and (1.9) in Eq. (1.7) to yield

$$(2G + \lambda)u_{,xx} + Gu_{,zz} + (G + \lambda)w_{,xz} = 0$$
$$(G + \lambda)u_{,xz} + (2G + \lambda)w_{,zz} + Gw_{,xx} = 0 \tag{1.10}$$

A solution for these equations can be chosen as

$$u = U(z) \cos\left(\frac{\pi x}{a}\right)$$
$$w = W(z) \sin\left(\frac{\pi x}{a}\right) \tag{1.11}$$

For this displacement field, σ_x will be of the form

$$\sigma_x = (..) \sin\left(\frac{\pi x}{a}\right) \tag{1.12}$$

It can be seen that the edge conditions (Eq. 1.5) are thus satisfied exactly. Substitution of Eq. (1.11) into Eq. (1.10) reduces the problem to the solution of the following coupled ordinary differential equations with constant coefficients:

$$-(2G+\lambda)\beta^2 U + GU'' + (G+\lambda)\beta W' = 0$$
$$-(G+\lambda)\beta U' + (2G+\lambda)W'' - G\beta^2 W = 0 \tag{1.13}$$

where $\beta = \frac{\pi}{a}$, and $(..)' = \frac{d(..)}{dz}$, $(..)'' = \frac{d^2(..)}{dz^2}$.

The solution of the above equations is straightforward and can be obtained as

$$U = A \cosh \beta z + B \sinh \beta z + Cz \cosh \beta z + Dz \sinh \beta z$$
$$W = \left[B - \frac{C(3-4\mu)}{\beta} \right] \cosh \beta z + \left[A - \frac{D(3-4\mu)}{\beta} \right] \sinh \beta z$$
$$+ Dz \cosh \beta z + Cz \sinh \beta z \tag{1.14}$$

where A, B, C and D are undetermined constants.

Rewriting the lateral surface conditions (Eq. 1.6) in terms of U, W, U', W', etc., one obtains four equations which are sufficient to evaluate the constants A to D. This procedure yields

$$A = \frac{D\left(1 - 2\mu - \frac{\beta h}{2} \coth \frac{\beta h}{2}\right)}{\beta}$$

$$B = \frac{C\left(1 - 2\mu - \frac{\beta h}{2} \tanh \frac{\beta h}{2}\right)}{\beta}$$

$$C = \frac{D \coth \frac{\beta h}{2}\left(\frac{\beta h}{2} + \sinh \frac{\beta h}{2} \cosh \frac{\beta h}{2}\right)}{\left(\frac{\beta h}{2} - \sinh \frac{\beta h}{2} \cosh \frac{\beta h}{2}\right)}$$

$$D = \frac{q_0(1+\mu)\sinh \frac{\beta h}{2}}{E(\beta h + \sinh \beta h)} \tag{1.15}$$

Use of Eqs. (1.11), (1.14) and (1.15) yields the values of u and w at any point of the domain. The strains and stresses are then obtained from Eqs. (1.8) and (1.9), respectively. Thus the solution of the problem is complete.

1.3   The Case of a Thin Plate

It is convenient to group the stresses and strains occurring in the structure into those corresponding to bending (σ_x, ε_x), transverse shear (τ_{xz}, γ_{xz}) and thickness stretch/contraction (σ_z, ε_z). The study of these quantities for various span-to-thickness ratios would enable one to arrive at the assumptions that form the basis of the theory of plates. Such a study has to be carried out by looking at the contributions of these stresses and strains to the strain energy of deformation of the structure. This is because any approximation to the three-dimensional problem can be considered acceptable only if it leads to a reasonably good estimate of the strain energy, and the contributions of the different stress and strain components to this energy provide a correct estimate of their relative importance.

The strain energy U_e for the present test problem can be written as

$$
\begin{aligned}
U_e &= \frac{1}{2} \int (\sigma_x \varepsilon_x + \tau_{xz} \gamma_{xz} + \sigma_z \varepsilon_z) dVol \\
&= \frac{1}{2} \int \sigma_x \varepsilon_x dVol + \frac{1}{2} \int \tau_{xz} \gamma_{xz} dVol + \frac{1}{2} \int \sigma_z \varepsilon_z dVol \\
&= U_b + U_s + U_t
\end{aligned}
\tag{1.16}
$$

where U_b is the strain energy corresponding to bending, U_s to transverse shear and U_t to thickness stretch/contraction.

By substituting for the stresses and strains, and integrating over the domain, one obtains

$$
\begin{aligned}
U_b = \frac{q_o^2 a^2}{384\pi \, E\Gamma} & [3(1 - \mu - 2\mu^2)(\sinh 4\beta h - 2 \sinh 2\beta h) \\
& + 12\beta h(1 - \mu - 2\mu^2)(\cosh 2\beta h - 1) \\
& - 12(\beta h)^2 (1 - \mu - 2\mu^2) \sinh 2\beta h \\
& + 8(\beta h)^3 \{-10 + 2\mu + 12\mu^2 + (7 + \mu - 6\mu^2) \cosh 2\beta h\} \\
& - 8(\beta h)^4 (1 + \mu) \sinh 2\beta h + 16(\beta h)^5 (1 + \mu)]
\end{aligned}
\tag{1.17}
$$

$$
\begin{aligned}
U_s = \frac{q_o^2 a^2 (1 + \mu)}{192\pi \, E\Gamma} & [(3 \sinh 4\beta h - 6 \sinh 2\beta h) + 12\beta h(\cosh 2\beta h - 1) \\
& - 12(\beta h)^2 \sinh 2\beta h - 8(\beta h)^3 (2 + \cosh 2\beta h) \\
& + 8(\beta h)^4 \sinh 2\beta h - 16(\beta h)^5]
\end{aligned}
\tag{1.18}
$$

$$U_t = \frac{q_o^2 a^2}{384\pi E\Gamma}[3(5 - \mu - 6\mu^2)(\sinh 4\beta h - 2\sinh 2\beta h)$$
$$+ 12\beta h(5 - \mu - 6\mu^2)(\cosh 2\beta h - 1)$$
$$- 12(\beta h)^2(5 - \mu - 6\mu^2)\sinh 2\beta h$$
$$- 8(\beta h)^3(5 - \mu - 6\mu^2)(2 + \cosh 2\beta h)$$
$$- 8(\beta h)^4(1 + \mu)\sinh 2\beta h + 16(1 + \mu)(\beta h)^5] \tag{1.19}$$

where $\Gamma = \left[\sinh^2 \beta h - (\beta h)^2\right]^2$.

A comparison of the relative magnitudes of these energies can be carried out as follows. Noting that βh $(=\pi h/a)$ is a small parameter for thin plates, the ratios (U_s/U_b) and (U_t/U_b) are written in the form of a power series in βh as

$$\frac{U_s}{U_b} = \frac{1}{5(1 - \mu)}(\beta h)^2 - \frac{(1 - 22\mu)}{1050(1 - \mu)^2}(\beta h)^4$$
$$- \frac{(11 - 4\mu - 196\mu^2)}{94{,}500(1 - \mu)^3}(\beta h)^6 + \text{etc.} \tag{1.20}$$

$$\frac{U_t}{U_b} = \frac{-\mu}{10(1 - \mu)}(\beta h)^2 + \frac{(65 - 129\mu + 43\mu^2)}{2100(1 - \mu)^2}(\beta h)^4$$
$$- \frac{(10 - 626\mu + 1204\mu^2 - 399\mu^3)}{189{,}000(1 - \mu)^3}(\beta h)^6 + \text{etc.} \tag{1.21}$$

For the limiting case of $\beta h \to 0$, it is clear that both (U_s/U_b) and (U_t/U_b) tend to zero. Thus, when the thickness is very small, the analysis can be simplified by approximating the stresses and strains such that U_s and U_t are zero, and such a structure can then be termed a 'thin plate'.

The energies U_s and U_t are made zero by assuming that σ_z, γ_{xz} and ε_z are all zero. (Strictly speaking, it is not necessary that both σ_z and ε_z be zero to make U_t zero; further, the transverse shear stress τ_{xz}, though directly proportional to γ_{xz}, cannot be neglected as its presence is required to satisfy the z-direction equilibrium for any element cut out of the transversely loaded plate. These will be discussed at length in Chap. 3). This assumption, that the transverse shear strain, the transverse normal strain and the corresponding normal stress are all negligible, forms the basis for the development of the classical plate theory as detailed out in the next chapter.

Summary

After a brief outline of the theory of elasticity, the problem of a simply supported rectangular strip under sinusoidally distributed transverse load has been solved rigorously. The strain energies corresponding to bending, transverse shear and transverse normal stretch/contraction have been compared. It has been shown for the case of a thin plate that the bending strain energy is predominant, and, in comparison with it, the other two energies tend to be negligibly small. On this basis, the neglect of the transverse shear strain, the transverse normal strain and the transverse normal stress has been identified as the main hypothesis for the development of the classical plate theory.

Conceptual Question

Refer to a book on Theory of Elasticity, and find out about the following:

(a)   the principle of complementary shear;
(b)   how λ and G can be expressed in terms of E and μ;
(c)   the two-dimensional case known as plane stress;
(d)   the strain energy of deformation for a general state of stress.

Chapter 2
Classical Plate Theory

A 'plate' is a thin, flat structural member subjected to loads that cause bending. Though, in reality, such a member is a three-dimensional body, analysis using the three-dimensional theory of elasticity is not essential if the thickness is small compared to the in-plane dimensions. By assuming reasonably realistic variations through the thickness, of displacements, strains and stresses, the problem can be reduced to one of two-dimensional analysis. Such a two-dimensional theory, most commonly employed for practical analysis, is the classical plate theory (CPT), also referred to as the thin plate theory. The theory leads to drastic simplification of the plate problem as compared to the elasticity approach, without unacceptable loss of accuracy.

Starting from the results obtained for a thin rectangular slab in the previous chapter, the assumptions of CPT are first enunciated here. Later, the theory is presented in a mathematical form, with reference to Cartesian coordinates, leading to a biharmonic governing equation. A clear exposition of the boundary conditions and the need for Kirchhoff's shear force is also included. Modifications arising due to an elastic foundation or thermal effects are briefly discussed. Finally, the expression for the strain energy of a deformed plate is derived.

© The Author(s) 2021
K. Bhaskar and T. K. Varadan, *Plates*,
https://doi.org/10.1007/978-3-030-69424-1_2

2.1 Assumptions of Classical Plate Theory

It should be kept in mind that the classical plate theory is a simplification of the three-dimensional problem to one of two dimensions—those in the plane of the plate. Hence, the purpose of the assumptions to be put forth here is to eliminate the thicknesswise coordinate from the final equations; in other words, if the displacements, strains and stresses can be found out at all points of a reference plane (say, the mid-plane of the plate), it should be possible, by virtue of the assumptions of the theory, to determine these quantities at any point of the three-dimensional structure.

The two assumptions that will be listed here were originally arrived at by intuitive reasoning and represent the culmination of progressive efforts of many illustrious scientists such as Sophie Germain, Lagrange, Navier, Poisson and Kirchhoff during the nineteenth century. Instead of tracing these historical developments, an alternative approach is adopted in this book, wherein the plate is looked upon as the limiting case of a rectangular slab and a study of the corresponding elasticity results would lead one to the assumptions of CPT. The advantage of this approach is three-fold:

(a)   One is exposed to the basics of the theory of elasticity and can easily appreciate the full essence of each assumption.
(b)   The relative complexity of three-dimensional analysis as compared to two-dimensional analysis, and hence the need of a simple plate theory, is made quite clear.
(c)   One can easily proceed on to the study of *shear deformation theories* which form an important part of the present state-of-the-art in plate analysis.

In the previous chapter, results of the elasticity approach were presented for an infinite strip undergoing cylindrical bending, and it was shown that in the limiting case of a thin plate, the transverse shear strain γ_{xz}, the transverse normal strain ε_z and the corresponding normal stress σ_z can be neglected. Extending the argument for the general case of bending of a finite plate referred to Cartesian x-y-z coordinates (x and y in-plane, and z transverse), the quantities that have to be neglected are γ_{xz}, γ_{yz}, ε_z and σ_z, and this is the fundamental starting point for the development of the classical plate theory applicable to thin plates.

In this chapter, we shall proceed further with the exposition of such a theory as applicable to a homogeneous plate made of an isotropic, linearly elastic material and undergoing small deformations due to transverse loads alone without any applied in-plane forces. Generalizations of this theory to account for heterogeneity (e.g. a layered plate), material orthotropy, large deflections and combined bending and stretching will be taken up later.

Using the strain–displacement relations (Eq. 1.2) for the neglected strain components, one has

$$\varepsilon_z = w_{,z} = 0 \tag{2.1a}$$

$$\gamma_{xz} = u_{,z} + w_{,x} = 0 \tag{2.1b}$$

$$\gamma_{yz} = v_{,z} + w_{,y} = 0 \tag{2.1c}$$

Equation (2.1a) implies that w is independent of z,

$$\text{i.e.} \quad w(x, y, z) = w(x, y) \tag{2.2a}$$

Using this, and integrating Eqs. (2.1b) and (2.1c) with respect to z, one obtains

$$u = -zw_{,x} + f(x, y) \tag{2.2b}$$

$$v = -zw_{,y} + g(x, y) \tag{2.2c}$$

where f and g are undetermined functions independent of z.

For small deformations (with the deflections of the plate small compared to the thickness) and for transverse loads alone without any applied in-plane edge loads, it is easy to visualize that points on the mid-plane of the plate tend to move predominantly in the transverse direction and their displacements in the in-plane directions would be negligible. Taking the origin of the z-coordinate at the mid-plane, one thus obtains

$$f(x, y) = g(x, y) = 0 \tag{2.3}$$

So, the displacement field of CPT can finally be written as

$$\begin{aligned} u(x, y, z) &= -zw_{,x} \\ v(x, y, z) &= -zw_{,y} \\ w(x, y, z) &= w(x, y) \end{aligned} \tag{2.4}$$

Physically, one can interpret the above displacement field as follows. With reference to Fig. 2.1, let attention be focussed on a normal AMB to the mid-plane of the undeformed plate. As the point M moves vertically to the position M′ due to a small deformation of the plate, and, as the mid-plane of the plate rotates through an angle $w_{,x}$ about the y-axis, the normal rotates through an angle θ_y and occupies the new position A′M′B′.

For small deformations, since $\sin\theta \cong \tan\theta \cong \theta$,

$$\theta_y = \frac{u_A}{h/2} = \frac{-(-h/2)w_{,x}}{h/2} = w_{,x} \tag{2.5}$$

Similarly, one can show that the rotation θ_x of the normal at any point about the x-axis is equal to $w_{,y}$.

Hence, the deformed normal is perpendicular to the deformed mid-surface of the plate. (One can also come to this conclusion directly from the fact that both γ_{xz} and γ_{yz} are neglected.) Combining this fact with the assumption of zero transverse

Fig. 2.1 Kinematics of deformation

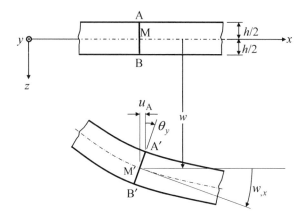

normal strain, one can physically interpret the displacement field of Eq. (2.4) in the form of the following statement, known as *Kirchhoff–Poisson hypothesis:*

Normals to the mid-plane of the undeformed plate suffer no change in length and remain normal to the deformed mid-surface.

Coming to stresses, enforcement of $\sigma_z = 0$ leads to a modification of the three-dimensional stress–strain law as indicated below.

$$\sigma_z = 2G\varepsilon_z + \lambda(\varepsilon_x + \varepsilon_y + \varepsilon_z) = 0 \tag{2.6}$$

i.e. $$\varepsilon_z = -\frac{\lambda}{(2G + \lambda)}(\varepsilon_x + \varepsilon_y) = -\frac{\mu}{(1 - \mu)}(\varepsilon_x + \varepsilon_y) \tag{2.7}$$

Thus, the equations

$$\sigma_x = 2G\varepsilon_x + \lambda(\varepsilon_x + \varepsilon_y + \varepsilon_z)$$
$$\sigma_y = 2G\varepsilon_y + \lambda(\varepsilon_x + \varepsilon_y + \varepsilon_z) \tag{2.8}$$

yield, after substitution of ε_z from Eq. (2.7) and simplification,

$$\sigma_x = \frac{E}{(1 - \mu^2)}(\varepsilon_x + \mu\varepsilon_y)$$
$$\sigma_y = \frac{E}{(1 - \mu^2)}(\varepsilon_y + \mu\varepsilon_x) \tag{2.9a}$$

The relation between the in-plane shear stress and strain is given by

$$\tau_{xy} = G\gamma_{xy} \tag{2.9b}$$

This stress–strain law, obtained by elimination of ε_z from the three-dimensional constitutive law by using $\sigma_z = 0$, is known as *plane-stress-reduced constitutive law*.

It should be kept in mind that $\varepsilon_z = 0$ is not enforced while deriving the above constitutive law. A careful look at the three-dimensional stress–strain relationships (Eq. 1.3) reveals that both σ_z and ε_z cannot be zero for a general state of stress. Thus, the assumption of negligible ε_z and that of negligible σ_z, which have been used as starting points to derive the plate theory, are mutually contradicting. Unless ε_z is taken to be zero, w would depend on z, and a simple displacement field as in Eq. (2.4) is not possible. Thus, from the viewpoint of obtaining a simple theory, the assumption of zero ε_z is necessary. The additional and contradictory assumption that $\sigma_z = 0$, resulting in the plane-stress-reduced constitutive law, is employed only because it leads to greater accuracy than the use of the unmodified three-dimensional constitutive law. This will be proved by means of numerical results in Sect. 3.3.

Another violation of the constitutive law occurs in CPT; this is with regard to the relation between the transverse shear strains γ_{xz} and γ_{yz}, and the corresponding shear stresses. A direct use of the constitutive law leads to zero τ_{xz} and τ_{yz} — a result not acceptable because they are required to counteract the applied transverse load. Thus, one has to assume the existence of non-zero transverse shear stresses; their actual determination, from purely equilibrium considerations, will be discussed in Sect. 2.6.

To sum up, the second assumption of CPT, besides Kirchhoff–Poisson hypothesis, can be stated as:

A plane-stress reduced constitutive law, obtained by neglecting the transverse normal stress, holds good. Further, while the transverse shear strains are taken as zero, the corresponding shear stresses do exist for the sake of ensuring equilibrium.

2.2 Moment–Curvature Relations

The in-plane strains corresponding to the displacement field of Eq. (2.4) are

$$\varepsilon_x = u_{,x} = -zw_{,xx}$$
$$\varepsilon_y = v_{,y} = -zw_{,yy}$$
$$\gamma_{xy} = u_{,y} + v_{,x} = -2zw_{,xy} \tag{2.10}$$

The quantities $-w_{,xx}$ and $-w_{,yy}$ are the bending curvatures in x and y directions, respectively, and $-2w_{,xy}$ is the twist curvature.

Use of the plane-stress-reduced constitutive law (Eq. 2.9) leads to

$$\sigma_x = -\frac{Ez}{(1-\mu^2)}(w_{,xx} + \mu w_{,yy})$$
$$\sigma_y = -\frac{Ez}{(1-\mu^2)}(w_{,yy} + \mu w_{,xx})$$
$$\tau_{xy} = -2Gzw_{,xy} = -\frac{E}{(1+\mu)}zw_{,xy} \tag{2.11}$$

Thus, the thicknesswise variation of the stresses and strains is linear, and anti-symmetric with respect to the mid-plane ($z = 0$). The mid-plane is free of stress and hence referred to as the *neutral surface.*

For a two-dimensional plate theory, the z-coordinate should not appear in the final equations and hence one cannot continue to employ the above z-dependent stresses; instead, one has to think in terms of analogous quantities applicable for the entire thickness of the plate. This is achieved by using stress resultants which are suitably chosen z-integrals of the stresses. Due to the antisymmetric variation of σ_x, σ_y and τ_{xy} with respect to z, the resulting action of these stresses is to cause bending and twisting of the plate. Thus, the appropriate stress resultants for the plate theory are bending and twisting moments per unit length as defined below:

$$(M_x, M_y, M_{xy}) = \int_{-h/2}^{h/2} (\sigma_x, \sigma_y, \tau_{xy})z \, dz \qquad (2.12)$$

The sign convention for these moments is as indicated in Fig. 2.2 where positive quantities are shown; these are consistent with the sign convention used for the stresses (see Fig. 1.1) and the absence of any minus signs in Eq. (2.12). It should be noted that M_x, M_y and M_{xy} are all functions only of x and y and are referred to the mid-plane of the plate. Further, at any point, M_{xy} will be acting simultaneously about the orthogonal x and y directions due to the principle of complementary shear.

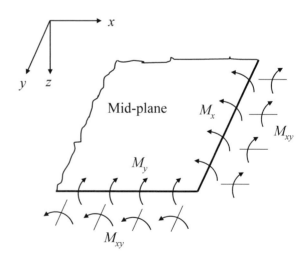

Fig. 2.2 Sign convention—positive moments

Substitution for the stresses from Eq. (2.11) into Eq. (2.12) yields

$$M_x = -D(w_{,xx} + \mu w_{,yy})$$

$$M_y = -D(w_{,yy} + \mu w_{,xx})$$

$$M_{xy} = -D(1 - \mu)w_{,xy} \tag{2.13}$$

where D defined as $\frac{Eh^3}{12(1-\mu^2)}$ is known as the flexural rigidity of the plate; it has the units of Nm.

Equations (2.13) give the *moment–curvature relations*; these should be looked upon as the constitutive law of CPT relating the generalized forces (M_x, M_y, M_{xy}) and the generalized displacements $(-w_{,xx}, -w_{,yy}, -2w_{,xy})$.

2.3 Equilibrium Equations

Consider a plate of arbitrary geometry and arbitrary boundary conditions. The plate is subjected to transverse load whose intensity q per unit area is a function of x and y; q is taken to be positive when acting in the positive z-direction. Let attention be focussed on a small rectangular element on the mid-plane of the plate, with its edges parallel to the x, y axes as shown in Fig. 2.3. As a free body, the element would be subjected to M_x, M_y and M_{xy} along the edges (shown by double arrows using the right-hand screw rule), and to the applied transverse load. This load can be assumed to be uniformly distributed over the small element, and its intensity can be taken as the value of $q(x,y)$ at the centre of the element. To balance the transverse load, shear resultants Q_x and Q_y per unit length, corresponding to the transverse shear stresses τ_{xz} and τ_{yz}, respectively, have to be introduced at the edges as shown. All the resultants shown in the figure are positive quantities as per the sign convention adopted here.

The moments M_x and M_{xy} change as one moves right through a distance dx in the x-direction. The corresponding values for the right edge can be obtained by a truncated Taylor expansion, acceptable since dx is small, as $(M_x + M_{x,x}\,dx)$ and $(M_{xy} + M_{xy,x}\,dx)$, respectively; similarly, Q_x changes to $(Q_x + Q_{x,x}\,dx)$. These incremented quantities are denoted by using a single star in Fig. 2.3. The incremented quantities in the y-direction are $(M_y + M_{y,y}\,dy)$, $(M_{xy} + M_{xy,y}\,dy)$ and $(Q_y + Q_{y,y}\,dy)$, denoted by using a double star.

Considering the equilibrium of forces and moments acting on the element and keeping in mind that all these resultants are defined per unit length of an edge, one obtains the following equations:

Fig. 2.3 Equilibrium of a
plate element

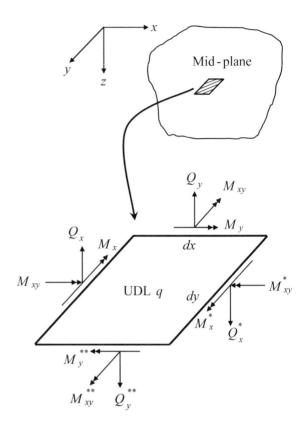

Force equilibrium in the z-direction

$$Q_{x,x}dx\,dy + Q_{y,y}dx\,dy + q\,dx\,dy = 0$$

i.e. $$Q_{x,x} + Q_{y,y} + q = 0 \qquad (2.14)$$

Moment equilibrium about a line parallel to the x-axis and passing through the centre of the element

$$-M_{xy,x}dx\,dy - M_{y,y}dx\,dy + (2Q_y + Q_{y,y}dy)dx\,\frac{dy}{2} = 0$$

Neglecting the term involving a product of three elemental lengths as small compared to the other terms, and simplifying, one gets

$$M_{xy,x} + M_{y,y} - Q_y = 0 \qquad (2.15)$$

Moment equilibrium about a line parallel to the y-axis and passing through the centre of the element

$$M_{x,x} + M_{xy,y} - Q_x = 0 \tag{2.16}$$

obtained in the same way as Eq. (2.15).

Equations (2.14)–(2.16) are the *equilibrium equations* of CPT and have to be satisfied at every point of the plate.

2.4 Governing Biharmonic Equation

Using Eqs. (2.15) and (2.16) to eliminate Q_x and Q_y from Eq. (2.14), one obtains

$$M_{x,xx} + 2M_{xy,xy} + M_{y,yy} + q = 0 \tag{2.17}$$

Substitution for M_x, M_y and M_{xy} in terms of the curvatures leads to

$$D(w_{,xxxx} + 2w_{,xxyy} + w_{,yyyy}) \equiv D\nabla^4 w = q \tag{2.18}$$

where $\nabla^4 \equiv \nabla^2.\nabla^2 \equiv \left(\frac{\partial^2}{\partial x^2} + \frac{\partial^2}{\partial y^2}\right).\left(\frac{\partial^2}{\partial x^2} + \frac{\partial^2}{\partial y^2}\right)$ is called the biharmonic operator. Equation (2.18) is called the biharmonic equation; it is the *governing equation* of the plate as per CPT.

2.5 Boundary Conditions

The governing equation is a fourth-order partial differential equation in x and y, and hence two boundary conditions have to be specified along any edge. The commonly considered classical boundary conditions are (see Fig. 2.4 for their pictorial representation):

(a) **Simply supported edge**

$$\text{At } x = \text{constant}: \quad w = M_x = 0$$

$$\text{i.e. } w = -D(w_{,xx} + \mu w_{,yy}) = 0$$

$$\text{i.e. } w = w_{,xx} = 0 \tag{2.19a}$$

since $w_{,yy}$ is automatically zero once $w = 0$ is enforced along the edge.

Fig. 2.4 Notation for boundary conditions

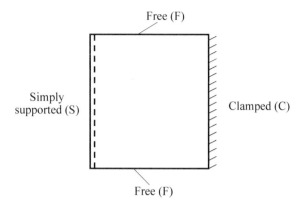

Free (F)

Simply supported (S)

Clamped (C)

Free (F)

Similarly,

$$\text{at } y = \text{constant: } w = w_{,yy} = 0 \qquad (2.19b)$$

(b)   Clamped edge

$$w = w_{,x} = 0 \quad \text{at } x = \text{constant}$$
$$w = w_{,y} = 0 \quad \text{at } y = \text{constant} \qquad (2.20)$$

(c)   Free edge

At a free edge, all the stress resultants corresponding to the normal and shear stresses on the edge plane have to be zero. Thus, strictly speaking, the boundary conditions at, say, $x = $ constant, have to be

$$M_x = M_{xy} = Q_x = 0 \qquad (2.21)$$

These conditions, known as Poisson's boundary conditions for a free edge, cannot be enforced simultaneously because only two conditions per edge can be satisfied by the fourth-order governing equation.

This difficulty is resolved by introducing a fictitious shear force called *Kirchhoff's shear force*. The moment M_{xy}, caused due to the action of horizontal shear stresses τ_{xy}, is imagined to be due to a distribution of vertical shear forces. Considering two adjacent elements of length dy each (Fig. 2.5), the corresponding twisting moments can be taken as $M_{xy} \, dy$ and $(M_{xy} + M_{xy,y} \, dy)dy$. These twisting moments are then imagined to be due to vertical shear forces acting in opposite directions at the ends of each element as shown. Considering many such elements and introducing fictitious shear forces for each element in the manner described above, one can see that the vertical shear force per unit length due to M_{xy} is given by $M_{xy,y}$ at any point of the edge. Thus the net vertical shear force per unit length would be $(Q_x + M_{xy,y})$; this quantity is known as Kirchhoff's shear force, denoted by V_x. By a similar argument,

Fig. 2.5 Kirchhoff's shear force

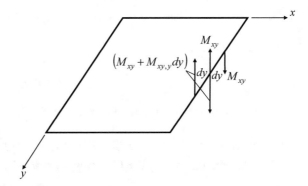

on the orthogonal edge, i.e. $y = $ constant, the Kirchhoff's shear force V_y per unit length is given by $(Q_y + M_{xy,x})$.

Thus, the free edge boundary conditions are prescribed as

$$M_x = V_x = 0 \text{ at } x = \text{constant}$$
$$M_y = V_y = 0 \text{ at } y = \text{constant} \tag{2.22}$$

These are called Kirchhoff's boundary conditions. It should be kept in mind that the simplification of the free edge conditions in the above form is an approximation, obtained by Kirchhoff using a variational derivation; it is necessary because of the inadequacy of the classical plate theory in that the intuitively correct free edge conditions as in Eq. (2.21) cannot be enforced.

If one has to express the conditions of Eq. (2.22) in terms of w, then expressions for Q_x and Q_y in terms of w are required. These are obtained using Eqs. (2.13), (2.15) and (2.16) as follows:

$$Q_x = M_{x,x} + M_{xy,y}$$
$$= [-D(w_{,xx} + \mu w_{,yy})]_{,x} + [-D(1 - \mu)w_{,xy}]_{,y}$$
$$= -D(w_{,xx} + w_{,yy})_{,x}$$

$$\text{i.e.} \quad Q_x = -D(\nabla^2 w)_{,x} \tag{2.23a}$$

Similarly,

$$Q_y = -D(\nabla^2 w)_{,y} \tag{2.23b}$$

Hence, using the definitions of V_x and V_y, one obtains

$$V_x = Q_x + M_{xy,y}$$
$$= -D(\nabla^2 w)_{,x} - D(1 - \mu)w_{,xyy}$$

$$\text{i.e.} \quad V_x = -D[w_{,xxx} + (2 - \mu)w_{,xyy}] \tag{2.24a}$$

Similarly,

$$V_y = -D[w_{,yyy} + (2 - \mu)w_{,xxy}] \tag{2.24b}$$

So, the free edge boundary conditions can finally be written as

$$w_{,xx} + \mu w_{,yy} = w_{,xxx} + (2 - \mu)w_{,xyy} = 0 \quad \text{at } x = \text{constant}$$
$$w_{,yy} + \mu w_{,xx} = w_{,yyy} + (2 - \mu)w_{,xxy} = 0 \quad \text{at } y = \text{constant} \tag{2.25}$$

At this stage, let us note that that the replacement of M_{xy} by a couple of vertical shear forces as done above leads not only to a continuously distributed shear force all along the edge, but also to concentrated forces at the corners. For the edge $x = \text{constant}$ (Fig. 2.5), these forces at the two corners are in opposite directions and of magnitude equal to the value of M_{xy} at the respective corner. An equal contribution to the corner force would arise from the orthogonal edge if it is also free. Thus, for a plate free at both the edges meeting a corner, one has to specify the corner force as zero,

$$\text{i.e.} \quad 2 M_{xy}\big|_{\text{corner}} = 0 \tag{2.26}$$

besides the free edge conditions for the two intersecting edges.

The above discussion is also of great relevance to the case of a rectangular plate with point supports only at the corners—a practical example being a plate supported by four columns. In this case, while the boundary condition at any corner can be specified easily as a completely restrained deflection, the corresponding support reaction itself has to be calculated as the local value of $2M_{xy}$ (see Sect. 4.4).

The case of an elastically restrained edge also needs to be discussed here. Let us consider an edge of the plate parallel to the y-axis and attached to another flexible member that is idealized as a set of torsional and linear springs with a rotational stiffness $k_t(y)$ and a translational stiffness $k(y)$, respectively. Thus, along the edge, the bending moment M_x is related to the normal slope $w_{,x}$ through the stiffness $k_t(y)$ as

$$M_x = \pm k_t w_{,x} \tag{2.27a}$$

with the correct sign such that both M_x and $w_{,x}$ are in the same direction. Similarly, the deflection w along the edge is linearly related to the corresponding shear force; this shear force, as per Kirchhoff's variational derivation, has to be V_x and not Q_x. This can be explained by noting that a deflection of the edge can be caused not only by a distribution of Q_x but also by M_{xy} and hence depends on the combined equivalent force V_x. Thus, one has

$$V_x = \pm kw \tag{2.27b}$$

with the proper sign as appropriate.

Without going deeper into the issue at this stage, let us also discuss the edge reaction that restrains the deflections along a simply supported edge. By summing the shear forces Q_x and Q_y along the four edges of a rectangular plate simply supported all around (see Problem 3 of Chap. 4), one can show that they balance the total transverse load. However, as per the variational derivation, the force corresponding to the transverse deflection along an edge is Kirchhoff's shear force and not the ordinary shear force, and this has to be valid for a simply supported edge as well; as a justification of this statement, one can show that the total load of the simply supported plate is also balanced by the shear forces V_x and V_y along with the four corner forces. The question whether the shear forces Q_x and Q_y without corner forces, or Kirchhoff's shear forces V_x and V_y along with the corner forces, have to be taken as the correct support reactions of the simply supported plate, can be answered satisfactorily only when a more complete description of the simple support is available. This will be taken up later in Sect. 10.5 when the problem is analysed using the theory of elasticity.

2.6   Solution of a Problem

Once the loading and the support description are given, the solution for the problem of a plate can be sought in terms of the deflection function $w(x,y)$ satisfying the governing biharmonic equation and the boundary conditions. The stress resultants M_x, M_y, M_{xy}, Q_x and Q_y at any point of the plate can then be determined using Eqs. (2.13) and (2.23).

To determine the maximum in-plane bending and shear stresses σ_x, σ_y and τ_{xy}, which occur at $z = \pm h/2$, it is necessary to relate these values to M_x, M_y and M_{xy}. This is easily done because, as per CPT, the distribution of these stresses with respect to z is linear and antisymmetric about the mid-plane. The resulting equation is

$$(\sigma_x, \sigma_y, \tau_{xy})_{\text{at } z=\pm\frac{h}{2}} = \pm\frac{6}{h^2}(M_x, M_y, M_{xy}) \tag{2.28}$$

The determination of the maximum transverse shear stresses τ_{xz} and τ_{yz} is not straightforward. It should be recollected that the shear strains γ_{xz} and γ_{yz} are assumed to be zero in CPT, and hence the corresponding shear stresses take the unrealistic value of zero as per the constitutive law. (These zero values are known as the *direct* estimates of the shear stresses). However, from equilibrium considerations, the shear forces Q_x and Q_y are non-zero, and hence there is a possibility of obtaining the shear stresses from these forces. (Such estimates are called *statically equivalent* estimates.) For this purpose, it is necessary to find out how the shear stresses vary through the thickness.

A careful look at the three-dimensional equilibrium equations reveals that a parabolic variation of τ_{xz} and τ_{yz} through the thickness is consistent with a linear variation of σ_x, σ_y and τ_{xy}. For example, the first equilibrium equation can be rewritten as

$$\tau_{xz,z} = -(\sigma_{x,x} + \tau_{xy,y})$$

$$\text{i.e.}\quad \tau_{xz}\Big|_{-h/2}^{z} = -\int_{-h/2}^{z} (\sigma_{x,x} + \tau_{xy,y})dz$$

$$\text{i.e.}\quad \tau_{xz} = -\int_{-h/2}^{z} (\sigma_{x,x} + \tau_{xy,y})dz \quad \text{at any } z \qquad (2.29)$$

since $\tau_{xz} = 0$ at $z = -h/2$. The integrand on the right-hand side is a linear (and odd) function of z, and hence the integral turns out to be a quadratic (and even) function of z.

Thus, from equilibrium considerations, τ_{xz} and similarly τ_{yz} vary parabolically through the thickness with zero values at $z = \pm h/2$, as given by

$$(\tau_{xz}, \tau_{yz}) \quad \propto \quad \left(z^2 - \frac{h^2}{4}\right) \qquad (2.30)$$

Equating the integrals of these over the thickness to Q_x and Q_y, respectively, one can obtain the maximum transverse shear stresses as

$$(\tau_{xz}, \tau_{yz})_{\text{at } z=0} = \frac{3}{2h}(Q_x, Q_y) \qquad (2.31)$$

It should be noted that there is a violation of the three-dimensional constitutive law in the above approach, i.e. the above non-zero transverse shear stresses are associated with zero transverse shear strains.

2.7 Inclusion of an Elastic Foundation/Thermal Effects

Two complicating effects which can be easily accounted for in the above theory are the presence of an elastic foundation and the effect of temperature variations. These are briefly discussed below.

Fig. 2.6 Winkler foundation

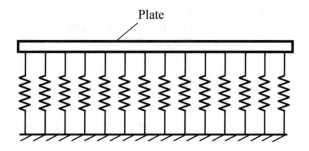

2.7.1 Elastic Foundation

An elastic foundation is of relevance not only to problems such as the analysis of a pavement supported by soil, but also in other cases such as the analysis of thin films deposited on a subgrade. The simplest and most commonly considered foundation model is known as Winkler's model wherein the foundation is replaced by a set of identical but mutually independent, continuously distributed linear springs (Fig. 2.6). Thus, the only reaction from the foundation to the plate is a lateral pressure $p(x,y)$ related to the deflection $w(x,y)$ of the plate as

$$p = -k_F w \qquad (2.32)$$

where k_F is the foundation stiffness or modulus, with units of N/m^3.

Adding this to the transverse load $q(x,y)$, the governing equation for the plate on such a foundation becomes

$$D\nabla^4 w + k_F w = q \qquad (2.33)$$

This has to be solved along with appropriate edge conditions which remain the same as before and unaffected by the presence of the foundation.

2.7.2 Thermal Effects

Temperature variations are commonly encountered and often need to be accounted for. Here we shall limit our discussion to plates subjected to temperature changes that are antisymmetric with respect to the mid-plane,

$$\text{i.e.} \qquad \Delta T(x, y, -z) = -\Delta T(x, y, z) \qquad (2.34)$$

so that they cause bending without mid-plane stretching.

The thermoelastic thin plate problem is similar to the isothermal case and based on the same assumptions except for a change in the plane-stress-reduced constitutive law to include linear thermal expansions. Starting with the expressions for the strains as

$$\varepsilon_x = \frac{\sigma_x - \mu\sigma_y}{E} + \alpha\Delta T$$

$$\varepsilon_y = \frac{\sigma_y - \mu\sigma_x}{E} + \alpha\Delta T \tag{2.35}$$

where α is the coefficient of thermal expansion, and inverting them, one gets

$$\sigma_x = \frac{E}{(1-\mu^2)}\left[\varepsilon_x + \mu\varepsilon_y - (1+\mu)\alpha\Delta T\right]$$

$$\sigma_y = \frac{E}{(1-\mu^2)}\left[\varepsilon_y + \mu\varepsilon_x - (1+\mu)\alpha\Delta T\right] \tag{2.36}$$

in place of Eq. (2.9a). The relation between τ_{xy} and γ_{xy} remains unaffected.

Proceeding further as before, the moment–curvature relations become

$$M_x = -D(w_{,xx} + \mu w_{,yy}) - M_T$$

$$M_y = -D(w_{,yy} + \mu w_{,xx}) - M_T$$

$$M_{xy} = -D(1-\mu)w_{,xy} \tag{2.37}$$

where $M_T = \int_{-h/2}^{h/2} \frac{E\alpha}{(1-\mu)}\Delta T\, z\mathrm{d}z$ is referred to as the thermal moment and is a function of x, y.

The equilibrium equations in terms of the stress resultants remain the same as before leading to Eq. (2.17). After substitution from Eq. (2.37), one finally gets

$$D\nabla^4 w = q - \nabla^2 M_T \tag{2.38}$$

as the governing equation for the present thermoelastic problem.

The edge conditions, in terms of the deflection, slope, bending moment and shear force, remain the same as discussed earlier, but when the moment and shear force conditions are expressed in terms of w, the thermal term has to be considered. For example, the zero M_x condition at a simply supported edge $x = $ constant becomes

$$-Dw_{,xx} - M_T = 0 \tag{2.39}$$

while the zero V_x condition, if the edge is free, becomes

$$-D[w_{,xxx} + (2-\mu)w_{,xyy}] - M_{T,x} = 0 \tag{2.40}$$

2.8 Strain Energy of the Plate

The strain energy of deformation, as per CPT, is actually the strain energy corresponding to bending since the other two components due to transverse shear and thickness stretch/contraction are taken as negligible. The strain energy U_e is given by

$$
\begin{aligned}
U_e &= \frac{1}{2} \int (\sigma_x \varepsilon_x + \sigma_y \varepsilon_y + \tau_{xy} \gamma_{xy}) dVol \\
&= \frac{1}{2} \int \left(-w_{,xx} \int_{-h/2}^{h/2} \sigma_x z\, dz - w_{,yy} \int_{-h/2}^{h/2} \sigma_y z\, dz - 2w_{,xy} \int_{-h/2}^{h/2} \tau_{xy} z\, dz \right) dArea \\
&= \frac{1}{2} \int [M_x(-w_{,xx}) + M_y(-w_{,yy}) + 2M_{xy}(-w_{,xy})]\, dArea
\end{aligned}
\tag{2.41}
$$

This equation can be interpreted as the work done by the (internal) generalized forces as the plate undergoes the corresponding generalized displacements.

By the use of the moment–curvature relations, the above equation can be rewritten in terms of w alone as

$$
U_e = \frac{D}{2} \int [(\nabla^2 w)^2 - 2(1-\mu)(w_{,xx} w_{,yy} - w_{,xy}^2)]\, dArea
\tag{2.42}
$$

Summary

Classical plate theory has been presented in this chapter with a lucid discussion of the basic assumptions. The displacement field, the plane-stress-reduced constitutive law, the moment–curvature relations, the equilibrium equations, the governing biharmonic equation and some commonly considered boundary conditions have been explained. The expression for the strain energy of the plate has also been derived.

Conceptual Questions

1. List the assumptions of the engineering theory of beams. Are you able to look at the thin plate theory as a natural extension of the theory of beams?
2. Write down all the equations of CPT as applicable to cylindrical bending. By comparing these with the equations of engineering beam theory, note that the bending behaviour of an infinite plate strip is exactly the same as that of the beam but for a change in flexural rigidity. (This fact is useful because the results available for bending, buckling and vibrations of beams in handbooks can be used directly to obtain corresponding results for the plate strip.)

3.   The boundary conditions for a simply supported or a clamped edge have been specified in this chapter in terms of only two equations while, for the free edge, three equations known as Poisson's boundary conditions have been specified to be actually necessary to correctly describe the support. Discuss why the difficulty of a third equation does not arise with a simple or a clamped support. (*Hint: Find out what happens to M_{xy} and v at the edge x=constant.*)

Problem

Consider a simply supported beam of length L, depth d and unit width. Let it be subjected to uniformly distributed load of intensity q_o on the top surface. Using the engineering theory of beams, obtain the maximum bending stress. The maximum transverse normal stress (neglected in the theory) actually occurs on the loaded surface and is equal to q_o. By comparing this value with that of the maximum bending stress, show that it is smaller by a factor proportional to $(d/L)^2$ and is hence negligible for a long, slender beam. (This result forms a basis for neglecting σ_z in CPT.)

Chapter 3
A Critical Assessment of Classical Plate Theory

The accuracy of the plate theory presented in the earlier chapter is critically examined here using the elasticity solution of Sect. 1.2 as a benchmark. The parameters for such an assessment are deflections, stresses and the strain energy due to bending. The necessity of the plane-stress reduced constitutive law is also brought forth.

3.1 CPT Solution for the Test Problem of Sect. 1.2

Let us recollect that the problem considered earlier was regarding the deformation of an infinitely long rectangular slab subjected to sinusoidal transverse loading on the top surface and supported by shear diaphragms at the longitudinal edges. For analysis using CPT, the slab should now be looked upon as a thin plate with the edges simply supported. Noting that $\partial(\ )/\partial y = 0$, the governing equation becomes

$$Dw_{,xxxx} = q_o \sin \frac{\pi x}{a} \tag{3.1}$$

with the following boundary conditions:

$$w = w_{,xx} = 0 \text{ at } x = 0, a \tag{3.2}$$

A solution for Eq. (3.1), satisfying the boundary conditions, can be taken as

$$w = W \sin \frac{\pi x}{a} \tag{3.3}$$

© The Author(s) 2021
K. Bhaskar and T. K. Varadan, *Plates*,
https://doi.org/10.1007/978-3-030-69424-1_3

(Alternatively, one can also solve Eq. (3.1) directly to obtain w as the sum of a complementary function and a particular integral; in that case, due to the boundary conditions, the complementary function will turn out to be zero).

By the use of Eq. (3.3) in Eq. (3.1), one gets

$$W = \frac{q_o}{D(\pi/a)^4} \tag{3.4}$$

The bending stress σ_x is then given by

$$\begin{aligned}
\sigma_x &= \frac{E}{(1-\mu^2)}\varepsilon_x \\
&= \frac{E}{(1-\mu^2)}(-zw_{,xx}) \\
&= \frac{Ez}{(1-\mu^2)}\left(\frac{\pi}{a}\right)^2 W \sin\frac{\pi x}{a} \tag{3.5}
\end{aligned}$$

The strain energy U_e per unit length of the infinite strip is given by

$$\begin{aligned}
U_e &= \frac{D}{2}\int_0^a (w_{,xx})^2 dx \\
&= \frac{\pi^4 D W^2}{4a^3} \tag{3.6}
\end{aligned}$$

3.2 Comparison with the Elasticity Solution

The maximum deflection, the maximum bending stress, and the total strain energy of the plate considered above are compared with the results of the elasticity solution for various values of a/h in Table 3.1. Results for the maximum transverse shear stress, obtained by using Eqs. (2.23a) and (2.31), are also included.

The parameters are non-dimensionalized such that CPT yields the same numerical value for all a/h, as given by

$$U_e^* = \left(\frac{E}{q_o}\right)\left(\frac{h}{a}\right)^3\left(\frac{U_e}{q_o a^2}\right) \quad w^* = \left(\frac{E}{q_o}\right)\left(\frac{h}{a}\right)^4\left(\frac{w}{h}\right)$$

$$\sigma_x^* = \left(\frac{h}{a}\right)^2\left(\frac{\sigma_x}{q_o}\right) \quad \tau_{xz}^* = \left(\frac{h}{a}\right)\left(\frac{\tau_{xz}}{q_o}\right)$$

From this table, the monotonic convergence of the CPT solution to the exact solution as the plate becomes thinner should first be noted. The strain energy, the

Table 3.1 Comparison of CPT and elasticity results

a/h	$U_e{}^*$		$w^*(a/2, 0)$		$\sigma_x^*\,(a/2, h/2)$		$\tau_{xz}^*\,(0,0)$	
	Elasticity	Error (%)	Elast.	Error (%)	Elast.	Error (%)	Elast.	Error (%)
5	0.0304	-7.7	0.123	-9.1	0.615	-1.2	0.476	0.3
10	0.0286	-1.9	0.115	-2.4	0.610	-0.3	0.477	≈ 0
25	0.0281	-0.3	0.113	-0.4	0.608	≈ 0	0.477	≈ 0
50	0.0280	≈ 0	0.112	≈ 0	0.608	≈ 0	0.477	≈ 0
CPT	0.0280		0.112		0.608		0.477	

Notes: μ is taken to be 0.3; Error $= \frac{\text{(CPT result}-\text{Elasticity result)}}{\text{Elasticity result}}$, shown as '$\approx 0$' when $<0.1\%$

maximum deflection and the maximum bending stress are always under-predicted by CPT; the maximum transverse shear stress is overpredicted but the error is extremely small. The error, for any a/h, is more for the deflection than for the other parameters. *From the results of this particular problem,* CPT can be considered to be very accurate for isotropic homogeneous plates if $a/h \geq 25$, and hence such plates can be categorized as '*thin*'.

At this juncture, a note of caution is appropriate. The parameter a, in a/h, should be looked upon as the half-wavelength of the applied sinusoidal load instead of as the total span of the plate. This is because the deformation of a simply supported plate of span $a/2$ with the load given by $q_o \sin \frac{\pi x}{a/2}$ would be identical to that of half of a simply supported plate of span a with the load given by $q_o \sin \frac{2\pi x}{a}$. By virtue of this argument, it is clear that the results of CPT turn worse as the half-wavelength of the applied sinusoidal load decreases. Coming to general loads, one has to think in terms of decomposition of the load into Fourier harmonics and the relative amplitudes of the different harmonics; once higher harmonics come into picture, the error of CPT would be higher than that for the single-half-wave loading considered here, for the same value of a/h. Thus, the error of CPT would be more if the load is uniformly distributed over the entire span instead of being a single sine-wave; for patch loads, the amplitudes associated with the higher harmonics increase as the load becomes more and more localized, and hence the error of CPT also grows, with the case of a concentrated point load being the most severe.

Secondly, if the edges were to be clamped in the above problem instead of being simply supported, by analogy with the elementary engineering solution of a clamped–clamped beam, the deformation (i.e. change of shape due to bending) of a central portion of the span between the lines of contraflexure would be the same as that of a simply supported plate of the same length. Thus, the equivalent simply supported span for a clamped plate, for use with Table 3.1 to determine the error of CPT, is smaller than the actual span. In other words, for the same span-to-thickness ratio, CPT would yield more accurate results for a simply supported plate than for a clamped plate. Generalizing this further, one can say that the accuracy of the CPT estimates decreases as the boundary conditions become stiffer.

It will be shown later (Chap. 10) that the accuracy of CPT also depends on the material properties if the plate was to be orthotropic or made up of layers of different materials.

Finally, it should be kept in mind that there are certain problems in which a three-dimensional state of stress prevails—such as those of stress concentration due to holes, notches and at locations directly under a concentrated load or near a point or line support. At these locations, the assumptions of the plate theory are not valid and one has to carry out a complete, most often local, three-dimensional stress analysis.

3.3   Why the Plane–stress Constitutive Law ?

It was shown in Sect. 1.3, on the basis of an elasticity solution, that the strain energy contribution of the transverse normal stress and strain is negligible for a thin plate. Since this energy depends on the product of σ_z and ε_z, it is actually sufficient to neglect either of these and not both as has been done in CPT; let us now see why both have to be neglected.

If one does not neglect ε_z, then the displacement field cannot be as simple as that of Eq. (2.4) and the resulting theory would be more complicated. Thus, continuing with the zero-ε_z assumption, one can attempt to develop an alternative to CPT without neglecting σ_z as well. This implies that, instead of the plane–stress reduced constitutive law of CPT, the three-dimensional stress–strain law would be used, i.e.

$$
\begin{aligned}
\sigma_x &= 2G\varepsilon_x + \lambda(\varepsilon_x + \varepsilon_y) \\
\sigma_y &= 2G\varepsilon_y + \lambda(\varepsilon_x + \varepsilon_y) \\
\sigma_z &= \lambda(\varepsilon_x + \varepsilon_y)
\end{aligned}
\tag{3.7}
$$

along with $\tau_{xy} = G\gamma_{xy}$.

Proceeding on lines similar to the development of CPT, one obtains

$$
\begin{aligned}
M_x &= -D_1[(1-\mu)w_{,xx} + \mu w_{,yy}] \\
M_y &= -D_1[(1-\mu)w_{,yy} + \mu w_{,xx}] \\
M_{xy} &= -D(1-\mu)w_{,xy}
\end{aligned}
\tag{3.8}
$$

where $D_1 = \frac{Eh^3}{12(1+\mu)(1-2\mu)} = \frac{(1-\mu)D}{(1-2\mu)}$.

The derivation of the equilibrium equations in terms of the stress resultants is exactly the same as that in CPT, and hence the simplified final equilibrium equation is again

$$
M_{x,xx} + 2M_{xy,xy} + M_{y,yy} + q = 0
\tag{3.9}
$$

Using Eq. (3.8) in the above equation, the governing equation becomes, after simplification,

$$\frac{(1-\mu)^2}{(1-2\mu)}D\nabla^4 w = q \tag{3.10}$$

Comparing this with the governing equation of CPT, one can see that the flexural rigidity of the plate increases by the factor $(1-\mu)^2/(1-2\mu)$, which is equal to 1.225 for $\mu = 0.3$. This means that the deflections by this new theory would be less than the corresponding CPT estimates by a factor of 1.225. Recollecting that CPT itself underestimates the actual deflections, it is clear that the new theory yields worse results. By calculation, one can prove that the deterioration in accuracy of the maximum stress and the strain energy estimates of the new theory, with respect to those of CPT, is much higher than the deterioration of the deflection estimate (see Problem 2).

Thus, between the plane-stress reduced constitutive law and the three-dimensional constitutive law, the former is better from the viewpoint of the accuracy of the final results.

Summary

The test problem of Sect. 1.2 has been solved here using CPT, and the results have been compared with the corresponding elasticity values. It has been shown that CPT yields accurate results for a 'thin' plate; a precise demarcation of what is 'thin' has been arrived at for this particular problem. All the same, the effects of localized loading, boundary conditions and stress concentration on the accuracy of CPT have been highlighted so that the difficulty in arriving at a universally applicable definition of a 'thin' plate is duly appreciated.

In addition, the need for the use of the contradicting assumptions of zero-ε_z and zero-σ_z in CPT has been explained. It has been shown that relaxation of either of these assumptions leads only to a more complicated or less accurate alternative to CPT.

Problems

1. Obtain the CPT solution for a clamped–clamped infinite strip subjected to single-half-wave sinusoidal transverse loading. By identifying the lines of contraflexure, find out the span-to-thickness ratio of the portion of the strip between them. How small is this compared to the overall span-to-thickness ratio? Does this result depend on the load distribution?
2. An alternative to CPT has been presented in Sect. 3.3 and the maximum deflection by this new theory, for the test problem considered here, has been shown to be less than the CPT value by a factor of 1.225. Carry out a similar comparison of the maximum bending stress and the strain energy.

Chapter 4
Analysis of Rectangular Plates

Solutions to the CPT equations are developed here for rectangular plates with certain boundary conditions. Only problems for which simple, exact analytical solutions are possible are discussed; approximate solutions form the topic of Chap. 8.

Navier's method, applicable for plates with all edges simply supported, is presented first. Some illustrative problems are worked out to demonstrate the simplicity of this method. Later, Levy's method, applicable for plates with two opposite edges simply supported, is explained with examples. A comparison of the convergence rate of the two methods is also given for the case of a plate simply supported all around. The chapter ends with a surprisingly simple solution for a problem involving corner supports.

4.1 Recapitulation of Fourier Series

Both Navier and Levy methods are based on expansion of the load and deflection functions in the form of Fourier series, and hence a brief look at Fourier series is appropriate at this stage.

If a periodic function $f(x)$ with period $2L$ is piecewise continuous in the interval $-L \leq x \leq L$ and has a left-hand derivative and a right-hand derivative at each point of that interval, then $f(x)$ can be expanded in a Fourier series as

$$f(x) = \frac{a_o}{2} + \sum_{n=1}^{\infty} \left(a_n \cos \frac{n\pi x}{L} + b_n \sin \frac{n\pi x}{L} \right) \tag{4.1}$$

where a_o, a_n and b_n, called the Fourier coefficients, are given by the Euler formulae

$$a_o = \frac{1}{L} \int_{-L}^{L} f(x)\,dx; \quad a_n = \frac{1}{L} \int_{-L}^{L} f(x) \cos \frac{n\pi x}{L}\,dx$$

© The Author(s) 2021
K. Bhaskar and T. K. Varadan, *Plates*,
https://doi.org/10.1007/978-3-030-69424-1_4

$$b_n = \frac{1}{L} \int\limits_{-L}^{L} f(x) \sin \frac{n\pi x}{L} \, dx \qquad (4.2)$$

The conditions imposed on $f(x)$ imply that it should be finite and should have only a finite number of discontinuities and a finite number of maxima and minima in the interval $(-L, L)$. At a discontinuity, the series converges to the mean of the left-hand and the right-hand limits of $f(x)$.

For problems where the domain is just $0 \le x \le L$ and the function $f(x)$ (corresponding to load, deflection, etc.) is given only in this interval, one can employ what are known as *half-range* series. Only the half-range sine series will be used in this chapter—it is obtained by first extending $f(x)$, defined in the interval $(0, L)$ (Fig. 4.1a), to an odd function in the interval $(-L, L)$, and then by making it periodic from $-\infty$ to ∞ (Fig. 4.1b).

For such a function, a_o and a_n as per Eq. (4.2) turn out to be zero, and the series is obtained as

$$f(x) = \sum_{n=1}^{\infty} b_n \sin \frac{n\pi x}{L} \qquad (4.3)$$

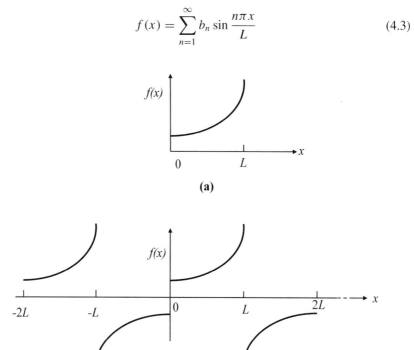

Fig. 4.1 Odd, periodic equivalent of a given function

where

$$b_n = \frac{1}{L} \int_{-L}^{L} f(x) \sin \frac{n\pi x}{L} dx = \frac{2}{L} \int_{0}^{L} f(x) \sin \frac{n\pi x}{L} dx \qquad (4.4)$$

The convergence of a Fourier expansion is usually very fast and depends on the continuity of the derivatives of $f(x)$. The Fourier coefficients decrease in magnitude towards zero, and this behaviour is governed by the following rule:

If $f(x)$ and its derivatives up to the order $(n-1)$ are continuous while the nth derivative has a finite number of isolated discontinuities, the convergence of the Fourier coefficients towards zero is like that of the sequence

$$1, \frac{1}{2^{n+1}}, \frac{1}{3^{n+1}}, \dots$$

Thus, a function which itself has a finite number of discontinuities has its Fourier coefficients decreasing like

$$1, \frac{1}{2}, \frac{1}{3}, \dots$$

For a continuous function whose first derivative has a finite number of discontinuities, the behaviour of the Fourier coefficients is like

$$1, \frac{1}{2^2}, \frac{1}{3^2}, \dots$$

Thus, for practical problems, assuming 'well-behaved' functions, it is usually sufficient to consider a truncated Fourier expansion with a small, finite number of terms.

It is also necessary to understand term-by-term differentiation of a Fourier series. If a function $f(x)$ is such that both $f(x)$ and $f'(x)$ satisfy the requirements for expansion as Fourier series, then the Fourier series for $f'(x)$ coincides with the term-by-term differentiation of the series for $f(x)$. This property is useful in two ways—to determine the Fourier expansions of the derivatives of a function defined originally as a Fourier series; and in the solution of differential equations where the unknown function is sought as a Fourier series and the nature of the variation (continuous, piecewise continuous, etc.) of the derivatives is known from a knowledge of the physics of the problem. It should always be kept in mind that the rate of convergence of the coefficients decreases as one differentiates a Fourier series.

4.2 Navier's Method

The analysis of a rectangular plate requires the solution of the governing biharmonic equation (Eq. (2.18)), along with appropriate boundary conditions at the four edges (Sect. 2.5), to yield the deflection function $w(x, y)$. Once this is carried out, the strains and stresses at any point of the plate structure can be obtained.

The basic idea behind Navier's method is to seek the solution for w in the form of an infinite series such that the edge conditions are satisfied *a priori* and the governing differential equation is reduced to simple algebraic equations. This is possible for a simply supported plate (Fig. 4.2) as detailed below.

It is assumed that the deflection function $w(x, y)$ is such that Fourier expansions exist for all its derivatives up to those occurring in the biharmonic equation, i.e. $w_{,xxxx}$, $w_{,yyyy}$ and $w_{,xxyy}$. This assumption is valid for most of the commonly encountered loads.

A look at the simply supported edge conditions given by

$$w = w_{,xx} = 0 \text{ at } x = 0, a$$
$$w = w_{,yy} = 0 \text{ at } y = 0, b \tag{4.5}$$

reveals that they can be satisfied *a priori* if the unknown deflection function is sought as a double sine series

Fig. 4.2 Simply supported rectangular plate

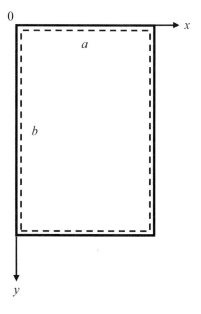

i.e.

$$w(x, y) = \sum_m \sum_n W_{mn} \sin \frac{m\pi x}{a} \sin \frac{n\pi y}{b} \qquad (4.6)$$

where W_{mn} are undetermined Fourier coefficients.

Substitution of this series for w in the governing equation yields

$$D\nabla^4 \left(\sum_m \sum_n W_{mn} \sin \frac{m\pi x}{a} \sin \frac{n\pi y}{b} \right) = q(x, y) \qquad (4.7)$$

Carrying out term-by-term differentiation, one obtains

$$D \sum_m \sum_n \left[\left(\frac{m\pi}{a} \right)^4 + 2 \left(\frac{m\pi}{a} \right)^2 \left(\frac{n\pi}{b} \right)^2 + \left(\frac{n\pi}{b} \right)^4 \right] W_{mn} \sin \frac{m\pi x}{a} \sin \frac{n\pi y}{b} = q \qquad (4.8)$$

Expanding q also in a double Fourier sine series as

$$q(x, y) = \sum_m \sum_n q_{mn} \sin \frac{m\pi x}{a} \sin \frac{n\pi y}{b} \qquad (4.9)$$

and noting that two Fourier series expansions can be equal only if all their corresponding coefficients are equal, one gets

$$D \left[\left(\frac{m\pi}{a} \right)^4 + 2 \left(\frac{m\pi}{a} \right)^2 \left(\frac{n\pi}{b} \right)^2 + \left(\frac{n\pi}{b} \right)^4 \right] W_{mn} = q_{mn}$$

i.e.

$$W_{mn} = \frac{q_{mn}}{D\pi^4 \left(\frac{m^2}{a^2} + \frac{n^2}{b^2} \right)^2} \qquad (4.10)$$

Hence,

$$w(x, y) = \sum_m \sum_n \frac{q_{mn}}{D\pi^4 \left(\frac{m^2}{a^2} + \frac{n^2}{b^2} \right)^2} \sin \frac{m\pi x}{a} \sin \frac{n\pi y}{b} \qquad (4.11)$$

Using the above solution for the deflection and the appropriate equations of Chap. 2, the strains, stresses, etc., at any point of the plate can be obtained. All these quantities will be in the form of infinite series, but as explained earlier, reasonably good estimates can be obtained by considering only the first few terms.

In the case of a plate on elastic foundation governed by Eq. (2.33), Navier's methodology is still applicable and leads to

$$W_{mn} = \frac{q_{mn}}{k_F + D\pi^4 \left(\frac{m^2}{a^2} + \frac{n^2}{b^2}\right)^2} \tag{4.12}$$

It is also applicable to the thermoelastic bending problem governed by Eq. (2.38) when the thermal moment is expanded in a double sine series as

$$M_T(x, y) = \sum_m \sum_n \psi_{mn} \sin\frac{m\pi x}{a} \sin\frac{n\pi y}{b} \tag{4.13}$$

For the case of a thermal moment alone without any mechanical loading, one gets

$$W_{mn} = \frac{\psi_{mn}}{D\pi^2 \left(\frac{m^2}{a^2} + \frac{n^2}{b^2}\right)} \tag{4.14}$$

Application of Navier's method is illustrated in the following examples.

Example 4.1: Uniform Load

Consider the case of a plate ($0 \le x \le a; 0 \le y \le b$) subjected to uniformly distributed load of intensity q_o per unit area.

The Fourier series corresponding to this uniform load is given by

$$q = \sum_{m=1,3,\,..}^{\infty} \sum_{n=1,3,\,..}^{\infty} \frac{16 q_o}{mn\pi^2} \sin\frac{m\pi x}{a} \sin\frac{n\pi y}{b}$$

wherein terms corresponding to even m or n do not appear in view of the symmetry of the load about the lines $x = a/2$ and $y = b/2$.

Using Eq. (4.11), the deflection function turns out to be

$$w = \sum_{m=1,3,\ldots}^{\infty} \sum_{n=1,3,\ldots}^{\infty} \frac{16 q_o}{D\pi^6 mn \left(\frac{m^2}{a^2} + \frac{n^2}{b^2}\right)^2} \sin\frac{m\pi x}{a} \sin\frac{n\pi y}{b}$$

Using this solution for w, the stress resultants at any point of the plate can be determined since they are all dependent on w or its derivatives. Finally, the maximum in-plane and transverse shear stresses can be calculated using Eqs. (2.28) and (2.31).

The convergence of the results, for the case of a square plate, is illustrated in Table 4.1. It should be noted that the rate of convergence of the series is very fast for w, while it is slower for M_x, M_y and M_{xy}, and further so for Q_x and Q_y. This is as can be expected because M_x, M_y and M_{xy} depend on the second derivatives of w while Q_x and Q_y depend on the third derivatives.

Table 4.1 Results by Navier's method for a square plate under uniformly distributed load

Terms	$100\left(\frac{Dw}{q_0 a^4}\right)$ at $\left(\frac{a}{2}, \frac{a}{2}\right)$	$10\left(\frac{M_x \text{ or } M_y}{q_0 a^2}\right)$ at $\left(\frac{a}{2}, \frac{a}{2}\right)$	$-10\left(\frac{M_{xy}}{q_0 a^2}\right)$ at $(0, 0)$	$\frac{Q_x(0, a/2)}{q_0 a}$ or $\frac{Q_y(a/2, 0)}{q_0 a}$
m, n up to 1	0.4161	0.5338	0.2875	0.2580
3	0.4055	0.4692	0.3140	0.2829
5	0.4064	0.4823	0.3198	0.3067
7	0.4062	0.4774	0.3220	0.3110
9	0.4062	0.4797	0.3230	0.3184
99	0.4062	0.4789	0.3248	0.3356
199	0.4062	0.4789	0.3248	0.3366

Notes: μ is taken to be 0.3; (x, y) locations are chosen to correspond to maximum deflection or stress resultant

The accuracy of the one-term approximation for $w_{\max}$ is very good (error with respect to the converged solution is 2.4%), while some more terms are required for obtaining reasonably accurate M_x, M_y, etc.

Example 4.2: Concentrated Load

As a second case, let us consider a square plate with a central concentrated load. One can solve this problem by considering it as the limiting case of a uniform load on a central patch of the plate as the patch dimensions tend to zero.

Let P be the magnitude of the concentrated load, and, to start with, let it be distributed uniformly over the patch of size $u \times u$ as shown in Fig. 4.3. Thus, $q = \frac{P}{u^2}$ in the region of the patch and zero elsewhere.

Fig. 4.3 A central patch load

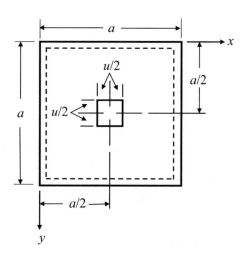

The Fourier coefficients corresponding to this load are

$$q_{mn} = \frac{4}{a^2} \int_{\frac{a-u}{2}}^{\frac{a+u}{2}} \int_{\frac{a-u}{2}}^{\frac{a+u}{2}} \frac{P}{u^2} \sin \frac{m\pi x}{a} \sin \frac{n\pi y}{a} \, dx \, dy$$

$$= \frac{4P}{a^2 u^2} \left[\frac{\cos \frac{m\pi(a-u)}{2a} - \cos \frac{m\pi(a+u)}{2a}}{\frac{m\pi}{a}} \right]$$

$$\left[\frac{\cos \frac{n\pi(a-u)}{2a} - \cos \frac{n\pi(a+u)}{2a}}{\frac{n\pi}{a}} \right]$$

$$= \frac{4P}{u^2 mn\pi^2} \left(2 \sin \frac{m\pi}{2} \sin \frac{m\pi u}{2a} \right) \left(2 \sin \frac{n\pi}{2} \sin \frac{n\pi u}{2a} \right)$$

$$= \begin{cases} \frac{16P}{u^2 mn\pi^2} (-1)^{\frac{m+n}{2}-1} \sin \frac{m\pi u}{2a} \sin \frac{n\pi u}{2a} & \text{for odd } m \text{ and } n \\ 0 & \text{if either } m \text{ or } n \text{ is even} \end{cases}$$

Hence,

$$w = \sum_{m=1,3,\dots}^{\infty} \sum_{n=1,3,\dots}^{\infty} \frac{16 P a^4 (-1)^{\frac{m+n}{2}-1}}{u^2 \pi^6 Dmn(m^2+n^2)^2} \sin \frac{m\pi u}{2a} \sin \frac{n\pi u}{2a} \sin \frac{m\pi x}{a} \sin \frac{n\pi y}{a}$$

The stress resultants corresponding to the patch load can be obtained by using the above solution for w.

Now, for the concentrated load, one has to take limits of these final equations as $u \to 0$. For example, noting that

$$\underset{u \to 0}{Lt} \frac{\sin \frac{m\pi u}{2a} \sin \frac{n\pi u}{2a}}{u^2} = \frac{mn\pi^2}{4a^2},$$

$$w = \sum_{m=1,3,\dots}^{\infty} \sum_{n=1,3,\dots}^{\infty} \frac{4 P a^2 (-1)^{\frac{m+n}{2}-1}}{\pi^4 D(m^2+n^2)^2} \sin \frac{m\pi x}{a} \sin \frac{n\pi y}{a}$$

This series can be shown to be convergent for all values of x and y, and the maximum deflection, at the centre, can be obtained reasonably accurately by considering only a small number of terms. (However, for any particular number of terms, the accuracy of the deflection for this case is less than that for the uniform load as obtained in Example 4.1. This is so because the deflection varies more steeply here and hence the corresponding Fourier coefficients decrease in magnitude more slowly).

The series for M_x, M_y, M_{xy}, Q_x and Q_y, obtained by taking limits as explained above, turn out to be convergent everywhere except at the point where the load is applied. At this particular point, all the stresses are singular (i.e. they tend to infinity) and hence these series are divergent. (In practice, loads, however localized they may be, are always applied over a small, yet non-zero, area, and hence the stresses directly below them are finite; further, due to local yielding, the size of the load patch increases and a redistribution of the high stresses takes place).

4.3 Levy's Method

This method is applicable for rectangular plates with two opposite edges simply supported, say, at $x = 0$ and $x = a$ (Fig. 4.4), and subjected to transverse loads which do not vary with respect to y; the boundary conditions at $y = \pm b/2$ can be arbitrary.

The basic idea is akin to that of Navier's method—instead of seeking a double sine series solution, here a single sine series in the x-direction is employed so that the simply supported edge conditions are satisfied *a priori* and the governing biharmonic equation is reduced to a fourth-order ordinary differential equation with constant coefficients. The four undetermined constants arising in the solution of this ordinary differential equation are obtained by enforcing the boundary conditions at the other two edges (at $y = \pm b/2$). The methodology is as detailed below.

The solution of the biharmonic equation

$$\nabla^4 w = \frac{q(x)}{D} \tag{4.15}$$

is sought in two parts as

Fig. 4.4 Coordinate system for Levy's method

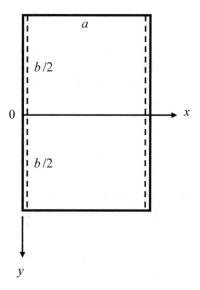

$$w(x, y) = w_1(x) + w_2(x, y) \tag{4.16}$$

The functions $w_1(x)$ and $w_2(x, y)$ are chosen such that they satisfy the following equations.

$$\nabla^4 w_1 = \frac{q(x)}{D} \tag{4.17}$$

and

$$\nabla^4 w_2 = 0 \tag{4.18}$$

i.e.

$$\nabla^4 (w_1 + w_2) = \frac{q(x)}{D} \tag{4.19}$$

The simply supported edge conditions at $x = 0, a$ are also split into two sets as

$$w_1 = w_{1,xx} = 0 \tag{4.20}$$

$$w_2 = w_{2,xx} = 0 \tag{4.21}$$

Equation (4.17) can be rewritten as

$$\frac{d^4 w_1}{dx^4} = \frac{q(x)}{D} \tag{4.22}$$

which, along with Eq. (4.20), yields a closed-form solution for w_1.

The solution for Eq. (4.18) is sought in the form of a Fourier sine series in x such that the conditions of Eq. (4.21) are automatically satisfied,

i.e.

$$w_2 = \sum_{m=1,2,\dots}^{\infty} W_m(y) \sin \frac{m \pi x}{a} \tag{4.23}$$

where W_m are undetermined functions of y.

Substitution of this series in Eq. (4.18) yields

$$\sum_{m=1,2,\dots}^{\infty} \left[W_{m,yyyy} - 2 \left(\frac{m\pi}{a} \right)^2 W_{m,yy} + \left(\frac{m\pi}{a} \right)^4 W_m \right] \sin \frac{m \pi x}{a} = 0 \tag{4.24}$$

i.e.

$$W_{m,yyyy} - 2\left(\frac{m\pi}{a}\right)^2 W_{m,yy} + \left(\frac{m\pi}{a}\right)^4 W_m = 0 \text{ for } m = 1, 2, \ldots \quad (4.25)$$

The solution for this homogeneous fourth-order ordinary differential equation can be written in terms of hyperbolic functions as

$$W_m = A_m \cosh \frac{m\pi y}{a} + B_m \frac{m\pi y}{a} \sinh \frac{m\pi y}{a}$$
$$+ C_m \sinh \frac{m\pi y}{a} + D_m \frac{m\pi y}{a} \cosh \frac{m\pi y}{a} \quad (4.26)$$

where A_m to D_m are constants. These constants are determined by using the boundary conditions at $y = \pm b/2$. This is best explained by means of an example as follows.

Example 4.3: Uniform Load on a Plate with all Sides Simply Supported
For $q(x) = q_0$, the solution for Eq. (4.22) along with the edge conditions of Eq. (4.20) can be found to be

$$w_1 = \frac{q_0}{24D}(x^4 - 2ax^3 + a^3 x)$$

which is symmetric with respect to the line $x = a/2$.

The total deflection $w_1 + w_2$, for the present problem, has to be symmetric about both the lines $x = a/2$ and $y = 0$. Since w_1 satisfies these symmetry requirements, they should be true for w_2 also. Thus, in Eq. (4.23), terms with even m do not appear; further, in Eq. (4.26), the constants C_m and D_m have to be zeroes. (In Fig. 4.4, the origin is taken at the mid-point of the left edge only to achieve this sort of simplification).

Thus, one gets

$$w_2 = \sum_{m=1,3,\ldots}^{\infty} \left(A_m \cosh \frac{m\pi y}{a} + B_m \frac{m\pi y}{a} \sinh \frac{m\pi y}{a} \right) \sin \frac{m\pi x}{a}$$

A_m and B_m are determined by using the edge conditions at $y = b/2$; the edge conditions at $y = -b/2$ will then get automatically satisfied because of the symmetry of the solution about the x-axis.

At $y = b/2$, for all x, we have

$$w = w_{,yy} = 0$$

i.e.

$$w_1 + w_2|_{y=b/2} = 0 \quad \text{and} \quad w_{2,yy}\big|_{y=b/2} = 0$$

At this stage, it is necessary to express w_1, which has been obtained in closed-form, as a sine series in x; as will be clear, this is to enable satisfaction of the first of the above conditions.

We have

$$w_1 = \frac{q_o}{24D}(x^4 - 2ax^3 + a^3x) = \frac{4q_oa^4}{\pi^5 D} \sum_{m=1,3,\dots} \frac{1}{m^5} \sin \frac{m\pi x}{a}$$

Hence, the first edge condition at $y = b/2$ yields

$$\sum_{m=1,3,\dots} \left(\frac{4q_oa^4}{\pi^5 Dm^5} + A_m \cosh \frac{m\pi b}{2a} + B_m \frac{m\pi b}{2a} \sinh \frac{m\pi b}{2a} \right) \sin \frac{m\pi x}{a} = 0$$

which leads to

$$\frac{4q_oa^4}{\pi^5 Dm^5} + A_m \cosh \frac{m\pi b}{2a} + B_m \frac{m\pi b}{2a} \sinh \frac{m\pi b}{2a} = 0$$

for each of the odd values of m.

Similarly, the second condition at $y = b/2$ leads to

$$(A_m + 2B_m) \cosh \frac{m\pi b}{2a} + B_m \frac{m\pi b}{2a} \sinh \frac{m\pi b}{2a} = 0$$

for each of the odd values of m.

Solving the above two equation sets, the constants A_m, B_m (for $m = 1, 3, \dots$) can be obtained as

$$A_m = \frac{-2q_oa^4(p_m \tanh p_m + 2)}{\pi^5 Dm^5 \cosh p_m}; \quad B_m = \frac{2q_oa^4}{\pi^5 Dm^5 \cosh p_m}$$

where

$$p_m = \frac{m\pi b}{2a}.$$

Table 4.2 Results by Levy's method for a uniformly loaded simply supported square plate

Terms	$100\left(\frac{Dw}{q_o a^4}\right)$ at $\left(\frac{a}{2},0\right)$	$10\left(\frac{M_x \text{ or } M_y}{q_o a^2}\right)$ at $\left(\frac{a}{2},0\right)$	$-10\left(\frac{M_{xy}}{q_o a^2}\right)$ at $\left(0,-\frac{a}{2}\right)$	$\frac{Q_x(0,0)}{q_o a}$ or $\frac{Q_y(a/2,-a/2)}{q_o a}$
m up to 1	0.4059	0.4766	0.3015	0.3385
3	0.4062	0.4789	0.3182	0.3377
5	0.4062	0.4789	0.3218	0.3377
13	0.4062	0.4789	0.3243	0.3377
39	0.4062	0.4789	0.3248	0.3377
69	0.4062	0.4789	0.3248	0.3377

Notes μ is taken to be 0.3
(x, y) locations are chosen to correspond to maximum deflection or stress resultant

The final solution is given by

$$w = \frac{q_o}{24D}(x^4 - 2ax^3 + a^3 x)$$
$$+ \sum_{m=1,3,\dots} \left(A_m \cosh\frac{m\pi y}{a} + B_m \frac{m\pi y}{a}\sinh\frac{m\pi y}{a}\right)\sin\frac{m\pi x}{a}$$

It should be noted that the series form of w_1 is not used here as that would lead to lesser accuracy in numerical calculations due to truncation error.

The results for a square plate, based on the above solution, are given in Table 4.2.

A comparison of the above results with those of Table 4.1 reveals the superior convergence of Levy's method compared to that of Navier's method; accurate estimates of both deflections and stresses can be obtained here by considering just the first few terms.

In the above example, symmetry of deformation was exploited to eliminate the even-m terms in Eq. (4.23) and the constants C_m and D_m. If, on the other hand, the load is unsymmetric about $x = a/2$, and/or the boundary conditions at $y = -b/2$ and those at $y = b/2$ are not identical, then such simplifications are not possible (see Problems 18–20).

Levy's method is also applicable for problems where some (or all) of the boundary conditions at $y = \pm b/2$ are non-homogeneous. This is illustrated in the following example, wherein the case of a simply supported plate undergoing flexure due to moments applied at the edges is considered.

Example 4.4: Simply Supported Plate Subjected to End Moments
The plate is free of transverse load but undergoes deformation due to externally applied end moments $M_1(x)$ and $M_2(x)$ as shown (Fig. 4.5).

For this case, w_1 turns out to be zero. The constants A_m to D_m are obtained using the following boundary conditions:

Fig. 4.5 End moments on a
simply supported plate

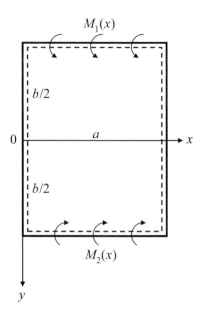

at $y = -b/2$:

$$\begin{cases} w = 0 \\ M_y = -D(w_{,yy} + \mu w_{,xx}) = -Dw_{,yy} = M_1(x) \end{cases}$$

at $y = b/2$:

$$\begin{cases} w = 0 \\ M_y = -D(w_{,yy} + \mu w_{,xx}) = -Dw_{,yy} = M_2(x) \end{cases}$$

These conditions can be rewritten as

at $y = -b/2$:

$$\begin{cases} \displaystyle\sum_{m=1,2,\dots} W_m \sin \frac{m\pi x}{a} = 0 \\ -D \displaystyle\sum_{m=1,2,\dots} W_{m,yy} \sin \frac{m\pi x}{a} = M_1(x) \end{cases}$$

at $y = b/2$:

$$\begin{cases} \displaystyle\sum_{m=1,2,\dots} W_m \sin \frac{m\pi x}{a} = 0 \\ -D \displaystyle\sum_{m=1,2,\dots} W_{m,yy} \sin \frac{m\pi x}{a} = M_2(x) \end{cases}$$

where W_m is given in Eq. (4.26).

It is clear that a simplification of the above equations is possible only by expressing $M_1(x)$ and $M_2(x)$ in the form of a sine series in x as

$$M_1(x) = \sum_m \alpha_m \sin \frac{m\pi x}{a}$$

$$M_2(x) = \sum_m \beta_m \sin \frac{m\pi x}{a}$$

Then, the edge conditions at $y = -b/2$ and $y = b/2$ reduce to

$$W_m|_{-b/2} = 0; \quad -DW_{m,yy}|_{-b/2} = \alpha_m$$

$$W_m|_{b/2} = 0; \quad -DW_{m,yy}|_{b/2} = \beta_m$$

for all values of m.

Substitution for W_m from Eq. (4.26) in the above conditions yields a set of four algebraic equations in A_m to D_m (for each m) which can be solved easily.

In case $M_1(x)$ and $M_2(x)$ are such that the deformation would be symmetric about either $x = a/2$ or $y = 0$ or both, the above equations can be simplified as was done in Example 4.3.

As explained earlier with reference to Navier's method, Levy's method may also be extended for plate problems involving an elastic foundation or thermal gradients which are invariant in the y-direction.

4.4 Closed-Form Solution for a Plate with Corner Supports

A surprisingly simple solution is possible for a rectangular plate resting on three corner supports and subjected to a point load at the fourth corner (Fig. 4.6)—a situation arising in practice when a plate is attached to four columns at its corners and one column is slightly taller or shorter than the others. The corresponding boundary

Fig. 4.6 Plate with three corner supports

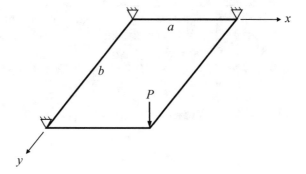

conditions are

$$
\begin{aligned}
&w = 0 && \text{at} (0, 0), (a, 0), (0, b); \\
&M_x = V_x = 0 && \text{at } x = 0, a; \\
&M_y = V_y = 0 && \text{at } y = 0, b; \\
&\text{Corner force} = P \text{ at} (a, b)
\end{aligned} \tag{4.27}
$$

The solution for this problem, obtained by simple trial and error, is

$$
w = Axy \tag{4.28}
$$

which satisfies the governing biharmonic equation and the homogeneous boundary conditions identically; the coefficient A has to be determined using the non-homogeneous corner force condition.

Noting that the corner force is given by $2M_{xy}$ as explained in Sect. 2.5, the corner condition can be expressed in terms of w as

$$
2D(1 - \mu)w_{,xy} = P \tag{4.29}
$$

where $w_{,xy}$ has to be positive for the downward force P. Using this, one gets

$$
A = \frac{P}{2D(1 - \mu)} \tag{4.30}
$$

Looking at the stress resultants, one can see that the bending moments are identically zero throughout the plate and the twisting moment is uniform and of value $P/2$. Thus, the plate is free of bending stresses and is subjected to shear stress τ_{xy} alone.

Summary

After a brief recapitulation of Fourier series, solution methodologies for the biharmonic equation in terms of double and single series, propounded originally by Navier and Levy, respectively, have been explained. A simple polynomial solution has also been presented.

It should be noted that the biharmonic equation is a linear equation and hence superposition is valid; i.e. the results for a combination of many loads can be obtained by simply summing the results corresponding to each load. This fact is very useful because important deflection and stress results for commonly encountered loads are listed out in various handbooks and can be combined to yield results for many other cases.

Conceptual Questions

1. Take a look at the result obtained for the maximum deflection of the uniformly
    loaded plate by Navier's method (Example 4.1) and answer the following.

    (i)   Find out how the deflection changes when
          (a)   the in-plane dimensions of the plate are doubled.
          (b)   the thickness of the plate is halved.
          (c)   the material is changed from mild steel to aluminium. (You should
                know that $E_{MS} \approx 3E_{Al}$; $\mu_{MS} \approx \mu_{Al} \approx 0.3$.)
    (ii) What is the central deflection corresponding to the loads given below?

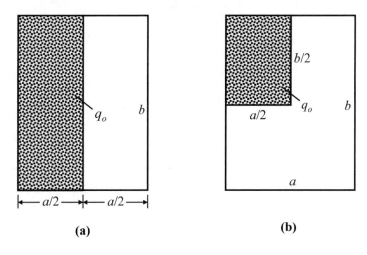

(a)   **(b)**

Question 1(ii)

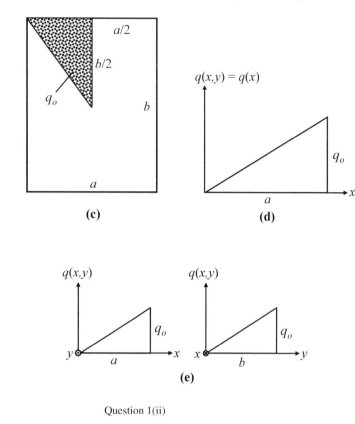

Question 1(ii)

2. You should know, from a first course on Mechanics of Materials, that there are three principal stresses σ_1, σ_2 and σ_3 at any point of a loaded body, and that the maximum shear stress is the largest of $\frac{|\sigma_1-\sigma_2|}{2}$, $\frac{|\sigma_2-\sigma_3|}{2}$ and $\frac{|\sigma_1-\sigma_3|}{2}$. Supposing one is required to design a simply supported square plate under uniform load on the basis of the maximum shear stress, give step-by-step instructions to find out this maximum stress. Be careful to consider both lateral surface and mid-surface locations.

Problems

Section 4.2

1. Express the loads shown in the form of double sine series.

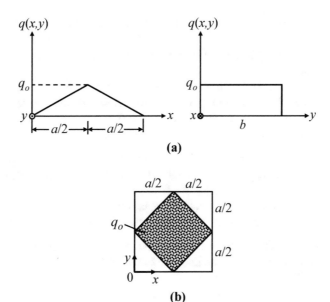

Problem 1

2. A simply supported square plate ($0 \leq x, y \leq a$) is subjected to uniform load over the area bounded by two squares of sides $a/4$ and $3a/4$. The sides of these squares are parallel to the edges of the plate and their centres coincide with the centre of the plate. Determine the deflection of the plate by Navier's method. (*Hint: The solution can be obtained very easily by superposition*).

3. For a simply supported plate ($0 \leq x \leq a$, $0 \leq y \leq b$) subjected to transverse pressure given by $q = q_o \sin \frac{\pi x}{a} \sin \frac{\pi y}{b}$, show that the total transverse load is balanced by the sum of the shear force reactions offered by the four edges.

4. Reproduce the results of Table 4.1 by actual calculation.

5. With reference to Example 4.1, taking terms up to m, $n = 9$, calculate values at a number of points to plot the following for a square plate:

   (a) w along $y = a/2$; (b) M_x along $y = a/2$;
   (c) M_{xy} along $x = 0$; (d) Q_x along $x = 0$.

6. A simply supported plate as in Fig. 4.2 is subjected to the transverse load given by

$$q = q_o \sin^2 \frac{\pi x}{a} \sin \frac{\pi y}{b}$$

   Find the solution for the deflection.

7. Consider a simply supported rectangular plate ($a = 2b$) subjected to transverse load distributed such that it is constant in the y-direction and varies linearly in the x-direction from zero at one end to q_o per unit area at the other end. For

this case, obtain the results for deflection and moments at the centre and shear forces at mid-edges, and show that the conclusions from Table 4.1 regarding the convergence of the Fourier series for these parameters hold good here also. Take $\mu = 0.3$.

8. A horizontal square plate 1 m $\times$ 1 m $\times$ 5 mm thick is simply supported at all edges. The plate is made of mild steel ($E = 200$ GPa, $\mu = 0.3$) and is subjected to a uniformly distributed load of 100 N/m² over a centrally located square patch of size 0.5 m $\times$ 0.5 m. For this case,

   (a) using a one-term Navier's approach, determine the maximum deflection (i) neglecting the weight of the plate; (ii) considering the weight of the plate. The density of MS is 7.8 gm/cc.

   (b) What will happen to the maximum deflection, due to the self-weight of the plate alone, if the thickness of the plate is changed to 4 mm?

9. (a) Examine the convergence of the series for w corresponding to Example 4.2 and compare this with that of Example 4.1.

   (b) Corresponding to Example 4.2, obtain the series for M_x and M_{xy}, and examine their convergence at ($a/4$, $a/4$).

10. (a) Obtain the deflection of the plate due to a transverse line load of P per unit length as shown.

   (b) Using the first four non-zero terms, determine the thickness required if the maximum deflection is not to exceed 1 mm. Take $E = 200$ GPa, $\mu = 0.3$, $a = b = 1$ m, $c = 0.8$ m and $P = 50$ N/m.

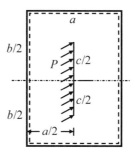

Problem 10

11. A square plate, 3 mm thick, is loaded as shown, P being the magnitude of each point load. Find P corresponding to a maximum deflection of 1 mm. Use a one-term Navier's approach. Take $E = 200$ GPa and $\mu = 0.3$.

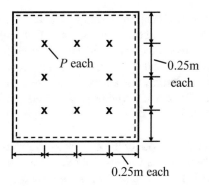

Problem 11

12. Using a one-term Navier's approach, determine the flexural rigidity required
     for the line-loaded plate shown if the maximum deflection is not to exceed
     5 mm. Repeat this if the total load were to be uniformly distributed over the
     plate.

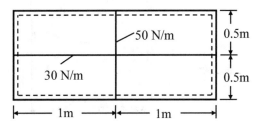

Problem 12

13. Consider a simply supported square plate subjected to $q = q_o \sin \frac{\pi x}{a} \sin \frac{\pi y}{a}$. If
     the maximum deflection is halved by providing an elastic foundation, what is
     its modulus?

Section 4.3

14. Reproduce the results of Table 4.2 by actual calculation.
15. (a) Obtain Levy's solution for the deflection of a plate, as shown in Fig. 4.4,
     subjected to transverse load given by $q = q_o \sin \frac{m\pi x}{a}$. The edges $x = 0$ and
     $x = a$ are simply supported while the other two edges are clamped.
     (b) Using this solution, determine the maximum deflection and maximum
     values of the stresses σ_x, σ_y and τ_{xy} if $a = b$, $\mu = 0.3$ and $m = 4$.

16. Repeat Problem 15(a) if the edges $y = \pm b/2$ were to be (a) simply supported; (b) free. Using the above solutions and that of Problem 15(a), and taking $m = 1$, show that the central deflections for all the three cases approach the same value as b/a tends to infinity. Explain why this is so.

17. Find an expression for the deflection of a uniformly loaded rectangular plate, as shown in Fig. 4.4, if the edges at $y = \pm b/2$ are free.

18. Repeat Problem 17 if the edge $y = b/2$ is clamped and the edge $y = -b/2$ is free.

19. Obtain Levy's solution for the deflection of a rectangular plate, as in Fig. 4.4. Assume all the edges to be simply supported and the load to be a uniform pressure on half the area of the plate bounded by $x = 0$ and $x = a/2$. Obtain the value of the maximum transverse shear stress τ_{xz}. (*Hint: See Problem 5*).

20. A square plate is subjected to transverse load varying linearly in one direction and constant in the other direction as shown. Determine the central deflection by using a three-term series. Take $E = 200$ GPa and $\mu = 0.3$.

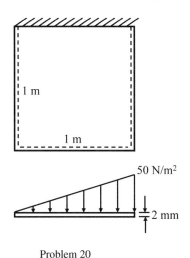

Problem 20

21. With reference to Example 4.4, determine the series for the deflection for:
     (a) $M_1(x) = M_2(x) = a$ constant M, (b) $M_1(x) = -M_2(x) = M$.

22.   Using the result of Problem 21, determine the central deflection of the plate
      shown. Take the first three non-zero terms. Take $E = 200$ GPa, $\mu = 0.3$, and
      the thickness to be 2.5 mm.

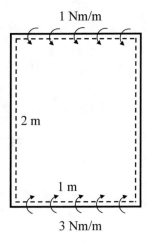

Problem 22

Chapter 5
Analysis of Circular Plates

The equations developed in Chaps. 1 and 2 with reference to Cartesian coordinates are not convenient for the analysis of circular plates. Hence, both the equations of the theory of elasticity and those of CPT are first given here with reference to cylindrical polar coordinates. The biharmonic governing equation for the circular plate is amenable to simple analytical treatment only for the special case of axisymmetric deformation. Such solutions are discussed here for a number of cases involving both solid and annular plates.

5.1 Equations of the Theory of Elasticity

With reference to r–θ–z coordinates as shown in Fig. 5.1, the field variables are the displacements along r, θ and z, designated respectively as u_r, u_θ and w, the strains $\varepsilon_r, \varepsilon_\theta, \varepsilon_z, \gamma_{\theta z}, \gamma_{rz}$ and $\gamma_{r\theta}$, and the stresses $\sigma_r, \sigma_\theta, \sigma_z, \tau_{\theta z}, \tau_{rz}$ and $\tau_{r\theta}$ acting on the six faces of the elemental volume.

In terms of these variables, the field equations of the theory of elasticity, applicable for the case without body forces, are:

Equilibrium Equations:

$$\sigma_{r,r} + \frac{\tau_{r\theta,\theta}}{r} + \tau_{rz,z} + \frac{(\sigma_r - \sigma_\theta)}{r} = 0$$
$$\tau_{r\theta,r} + \frac{\sigma_{\theta,\theta}}{r} + \tau_{\theta z,z} + \frac{2\tau_{r\theta}}{r} = 0$$

$$\tau_{rz,r} + \frac{\tau_{\theta z,\theta}}{r} + \sigma_{z,z} + \frac{\tau_{rz}}{r} = 0 \tag{5.1}$$

© The Author(s) 2021
K. Bhaskar and T. K. Varadan, *Plates*,
https://doi.org/10.1007/978-3-030-69424-1_5

Fig. 5.1 a Cylindrical polar coordinates. **b** An elemental volume

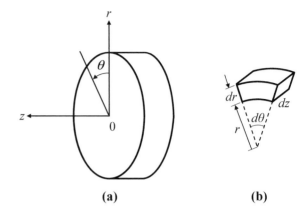

(a)   **(b)**

Strain–Displacement Relations:

$$\varepsilon_r = u_{r,r} \qquad \gamma_{\theta z} = u_{\theta,z} + \frac{w_{,\theta}}{r}$$

$$\varepsilon_\theta = \frac{(u_{\theta,\theta} + u_r)}{r} \qquad \gamma_{rz} = w_{,r} + u_{r,z}$$

$$\varepsilon_z = w_{,z} \qquad \gamma_{r\theta} = \frac{(u_{r,\theta} - u_\theta)}{r} + u_{\theta,r} \qquad (5.2)$$

Stress–Strain Law:

$$\sigma_i = 2G\varepsilon_i + \lambda e \quad \tau_{ij} = G\gamma_{ij} \text{ with } i, j = r, \theta, z \qquad (5.3)$$

where G and λ are Lame's constants, and e is the volumetric strain given by

$$e = \varepsilon_r + \varepsilon_\theta + \varepsilon_z$$

5.2   Equations of CPT

The displacement field, corresponding to zero transverse normal and shear strains (ε_z, $\gamma_{\theta z}$, γ_{rz}), is obtained as

$$u_r = -zw_{,r}$$

$$u_\theta = -z\frac{w_{,\theta}}{r}$$

$$w(r, \theta, z) = w(r, \theta) \qquad (5.4)$$

The in-plane strains are

$$\varepsilon_r = -zw_{,rr}$$

$$\varepsilon_\theta = -z\left(\frac{w_{,\theta\theta}}{r^2} + \frac{w_{,r}}{r}\right)$$

$$\gamma_{r\theta} = -2z\left(\frac{w_{,r\theta}}{r} - \frac{w_{,\theta}}{r^2}\right) \tag{5.5}$$

Assuming a plane-stress reduced constitutive law, the stresses are given by

$$\sigma_r = \frac{-Ez}{(1-\mu^2)}\left[w_{,rr} + \mu\left(\frac{w_{,\theta\theta}}{r^2} + \frac{w_{,r}}{r}\right)\right]$$

$$\sigma_\theta = \frac{-Ez}{(1-\mu^2)}\left[\frac{w_{,\theta\theta}}{r^2} + \frac{w_{,r}}{r} + \mu\, w_{,rr}\right]$$

$$\tau_{r\theta} = \frac{-Ez}{(1+\mu)}\left(\frac{w_{,r\theta}}{r} - \frac{w_{,\theta}}{r^2}\right) \tag{5.6}$$

The moments per unit length are defined as

$$(M_r, M_\theta, M_{r\theta}) = \int_{-h/2}^{h/2} (\sigma_r, \sigma_\theta, \tau_{r\theta})z\, dz \tag{5.7}$$

with the sign convention as shown in Fig. 5.2.

From Eqs. (5.6) and (5.7), one can obtain the following moment–curvature relations:

$$M_r = -D\left[w_{,rr} + \mu\left(\frac{w_{,\theta\theta}}{r^2} + \frac{w_{,r}}{r}\right)\right]$$

$$M_\theta = -D\left[\frac{w_{,\theta\theta}}{r^2} + \frac{w_{,r}}{r} + \mu\, w_{,rr}\right]$$

Fig. 5.2 Sign convention—positive moments

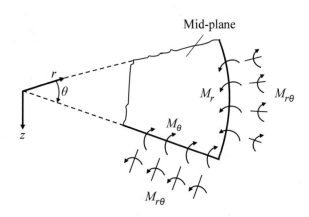

$$M_{r\theta} = -D(1 - \mu)\left(\frac{w_{,r\theta}}{r} - \frac{w_{,\theta}}{r^2}\right) \tag{5.8}$$

The equilibrium equations are obtained by considering the forces and moments on a small element of the plate. These are as shown in Fig. 5.3. It should be noted that Q_r and Q_θ, shear resultants per unit length, have to be introduced to balance the applied transverse load (exactly as was done in Chap. 2). The incremented quantities are denoted using the following relations:

$$(\ )^* = (\ ) + \frac{\partial \ (\ )}{\partial r}dr \qquad (\ )^{**} = (\ ) + \frac{\partial \ (\ )}{\partial \theta}d\theta$$

The summation of forces in the z-direction results in

$$-Q_r r \, d\theta + Q_r^*(r + dr)d\theta + (Q_\theta^{**} - Q_\theta)dr + q\left(r + \frac{dr}{2}\right)d\theta \, dr = 0$$

Fig. 5.3 Equilibrium of a plate element

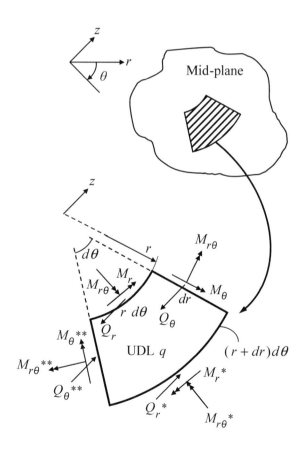

which yields, on simplification and neglect of the terms involving products of three small quantities,

$$Q_{r,r} + \frac{Q_{\theta,\theta}}{r} + \frac{Q_r}{r} + q = 0 \tag{5.9a}$$

Considering the moments about the θ-direction at the centre of the element, one gets

$$M_r^*(r + dr)d\theta - M_r r\, d\theta - Q_r^*(r + dr)d\theta\, \frac{dr}{2} - Q_r r\, d\theta\, \frac{dr}{2}$$
$$+ (M_{r\theta}^{**} - M_{r\theta})dr \cos \frac{d\theta}{2} - (M_\theta^{**} + M_\theta)dr \sin \frac{d\theta}{2} = 0$$

For small $d\theta$, $\sin \frac{d\theta}{2} \approx \frac{d\theta}{2}$ and $\cos \frac{d\theta}{2} \approx 1$, so that the above equation can be simplified as

$$M_{r,r} + \frac{M_{r\theta,\theta}}{r} + \frac{(M_r - M_\theta)}{r} - Q_r = 0 \tag{5.9b}$$

Similarly, by taking moments about the r-direction at the centre of the element, one obtains

$$M_{r\theta,r} + \frac{M_{\theta,\theta}}{r} + \frac{2M_{r\theta}}{r} - Q_\theta = 0 \tag{5.9c}$$

Use of Eqs. (5.9b) and (5.9c) reduces Eq. (5.9a) to

$$M_{r,rr} + \frac{2M_{r,r}}{r} - \frac{M_{\theta,r}}{r} + \frac{M_{\theta,\theta\theta}}{r^2} + \frac{2M_{r\theta,r\theta}}{r} + \frac{2M_{r\theta,\theta}}{r^2} + q = 0 \tag{5.10}$$

Using the moment–curvature relations, the above equation can be rewritten in terms of w to yield the biharmonic governing equation

$$\nabla^4 w = \frac{q}{D} \tag{5.11}$$

where

$$\nabla^4 \equiv \nabla^2.\nabla^2 \equiv \left(\frac{\partial^2}{\partial r^2} + \frac{1}{r}\frac{\partial}{\partial r} + \frac{1}{r^2}\frac{\partial^2}{\partial \theta^2} \right)\left(\frac{\partial^2}{\partial r^2} + \frac{1}{r}\frac{\partial}{\partial r} + \frac{1}{r^2}\frac{\partial^2}{\partial \theta^2} \right)$$
$$\equiv \frac{\partial^4}{\partial r^4} + \frac{2}{r}\frac{\partial^3}{\partial r^3} - \frac{1}{r^2}\frac{\partial^2}{\partial r^2} + \frac{1}{r^3}\frac{\partial}{\partial r} + \frac{2}{r^2}\frac{\partial^4}{\partial r^2\partial \theta^2}$$
$$- \frac{2}{r^3}\frac{\partial^3}{\partial r\partial \theta^2} + \frac{4}{r^4}\frac{\partial^2}{\partial \theta^2} + \frac{1}{r^4}\frac{\partial^4}{\partial \theta^4}$$

The boundary conditions are specified as:

$$w = w_{,r} = 0 \quad \text{for a clamped support}$$

$$w = M_r = 0 \quad \text{for a simple support}$$

$$M_r = Q_r + \frac{M_{r\theta,\theta}}{r} = 0 \quad \text{for a free boundary} \tag{5.12}$$

at the circumferential edges $r = \text{constant}$.

5.3 Solution of Axisymmetric Problems

The rigorous solution of Eq. (5.11) for a case of general deformation of a circular plate is rather complicated and beyond the scope of this book. For the special case of axisymmetric loading and axisymmetric boundary conditions, the governing equation gets simplified and becomes amenable to a straightforward solution, as explained below.

With $\frac{\partial(..)}{\partial \theta} = 0$, the governing equation becomes

$$\left(\frac{d^2}{dr^2} + \frac{1}{r}\frac{d}{dr} \right)\left(\frac{d^2 w}{dr^2} + \frac{1}{r}\frac{dw}{dr} \right) = \frac{q}{D} \tag{5.13}$$

which can be expressed more conveniently as

$$\frac{1}{r}\frac{d}{dr}\left[r\frac{d}{dr}\left\{ \frac{1}{r}\frac{d}{dr}\left(r\frac{dw}{dr} \right) \right\} \right] = \frac{q}{D} \tag{5.14}$$

When $q = q(r)$ is known, the above equation can be integrated successively to yield the deflection w. The four constants of integration are determined by enforcing the boundary conditions and the condition that the deflection and stress at the centre of a solid plate are finite. The procedure is illustrated in the following examples.

Example 5.1 Solid/annular plate under uniform load (Fig. 5.4).

Let $q(r) = q_o$ be the intensity of the uniform load. Successive integration of Eq. (5.14) yields

$$\frac{d}{dr}\left\{ \frac{1}{r}\frac{d}{dr}\left(r\frac{dw}{dr} \right) \right\} = \frac{q_o r}{2D} + \frac{c_1}{r}$$

$$\frac{d}{dr}\left(r\frac{dw}{dr} \right) = \frac{q_o r^3}{4D} + c_1 r \ln r + c_2 r$$

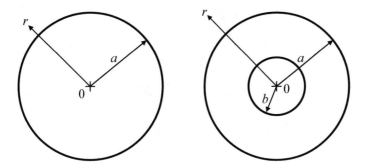

Fig. 5.4 Solid and annular plates

$$\frac{dw}{dr} = \frac{q_o r^3}{16D} + \frac{c_1}{2r}\left(r^2 \ln r - \frac{r^2}{2}\right) + \frac{c_2 r}{2} + \frac{c_3}{r}$$

$$w = \frac{q_o r^4}{64D} + \frac{c_1}{2}\left[\frac{1}{2}\left(r^2 \ln r - \frac{r^2}{2}\right) - \frac{r^2}{4}\right] + \frac{c_2 r^2}{4} + c_3 \ln r + c_4$$

where c_1 to c_4 are constants of integration. By a redefinition of these constants, the solution can be written as

$$w = \frac{q_o r^4}{64D} + \hat{c}_1 + \hat{c}_2 r^2 + \hat{c}_3 \ln r + \hat{c}_4 r^2 \ln r$$

It is more convenient if the argument of the logarithmic function is a normalized radius. This is achieved by a further redefinition of the constants to yield

$$w = \frac{q_o r^4}{64D} + C_1 + C_2 r^2 + C_3 \ln \frac{r}{a} + C_4 r^2 \ln \frac{r}{a}$$

For the solid plate, the deflection is finite at the centre ($r = 0$), and hence $C_3 = 0$. Similarly, it can be shown that C_4 has to be zero for the stress to be finite at the centre. C_1 and C_2 are obtained using the conditions at the periphery of the plate.

For the annular plate, there are four boundary conditions, two at the inner periphery and two at the outer periphery, and these are used to determine C_1 to C_4.

Once C_1 to C_4 are determined to yield the final expression for w, the moments, shear force, etc., can be determined by using the appropriate expressions for them in terms of w.

Example 5.2 A numerical problem.

Consider the design of a simply supported circular plate under uniform load. Given the load to be 100 kN/m^2 and the plate diameter to be 0.25 m, it is necessary to find out the thickness required such that the maximum deflection is limited to, say, 2 mm. The material is assumed to be aluminium with $E = 70$ GPa and $\mu = 0.3$. It is also necessary to find out the critical stresses.

Using the simply supported edge conditions

$$w = M_r = 0 \text{ at } r = a$$

i.e. $$w = w_{,rr} + \mu \frac{w_{,r}}{r} = 0 \text{ at } r = a$$

the constants C_1 and C_2 can be determined, yielding

$$w = \frac{q_o(a^2 - r^2)}{64D} \left(\frac{5 + \mu}{1 + \mu} a^2 - r^2 \right)$$

It is clear that the maximum deflection occurs at the centre and is given by

$$w_{max} = \left(\frac{5 + \mu}{1 + \mu} \right) \frac{q_o a^4}{64D}$$

For $a = 0.125$ m, $q_o = 100$ kN/m^2, $w_{max} = 2 \times 10^{-3}$ m and $\mu = 0.3$, one gets
$D = 777.61$ N m and hence $h_{required} = 0.00495$ m.

Coming to stresses, the maximum bending and shear stresses are obtained by
finding out the maximum values of M_r, M_θ and Q_r.

Using Eq. (5.8), one gets

$$M_r = -D\left(w_{,rr} + \frac{\mu w_{,r}}{r} \right) = (322.27 - 20625r^2)\text{N m/m}$$

$$M_\theta = -D\left(\frac{w_{,r}}{r} + \mu w_{,rr} \right) = (322.27 - 11875r^2)\text{N m/m}$$

where r is in metres.

It is clear that both M_r and M_θ are maximum at the centre of the plate and these
central values are equal; this is as it should be because, at the centre, any direction
can be looked upon as both radial and circumferential.

Thus, $M_{max} = 322.27$ N m/m and the corresponding bending stress is

$$\sigma_{max} = \frac{6M_{max}}{h^2} = 78.91 \text{ MPa}$$

Using Eq. (5.9b), we have

$$Q_r = M_{r,r} + \frac{(M_r - M_\theta)}{r} = -(50,000r) \text{ N/m}$$

Note that this can also be obtained by visualizing the force exerted at the periphery
of a circular portion of radius r to balance the lateral load acting on it (Fig. 5.5), thus
leading to

Fig. 5.5 Shear force due to uniform load

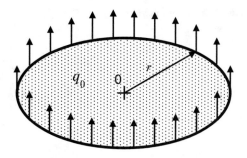

$$Q_r = \frac{-\pi\, r^2 q_0}{2\pi\, r} = \frac{-q_o r}{2} = -(50{,}000r)\ \text{N/m}$$

the negative sign being introduced as per the sign convention of Fig. 5.3.

$$\text{Thus,}\quad Q_{r\ \text{max}} = -(50{,}000a) = -6250\ \text{N/m}$$

$$\text{and hence,}\quad \tau_{rz\ \text{max}} = \frac{3}{2}\frac{Q_{r\ \text{max}}}{h} = 1.89\ \text{MPa}.$$

Example 5.3 Central concentrated load on a circular plate.

For this case, the governing equation [Eq. (5.14)] is not directly applicable. However, Q_r at any radius can be found out from equilibrium considerations as explained in Example 5.2, and this can be used to solve for w.
We have

$$Q_r = M_{r,r} + \frac{(M_r - M_\theta)}{r}$$

$$= -D\left(\frac{d^3w}{dr^3} + \frac{1}{r}\frac{d^2w}{dr^2} - \frac{1}{r^2}\frac{dw}{dr}\right)$$

$$= -D\frac{d}{dr}\left\{\frac{1}{r}\frac{d}{dr}\left(r\frac{dw}{dr}\right)\right\}$$

Thus,

$$\frac{d}{dr}\left\{\frac{1}{r}\frac{d}{dr}\left(r\frac{dw}{dr}\right)\right\} = -\frac{Q_r}{D}$$

For a central concentrated load P, at any radius r, from Fig. 5.6,

$$-(2\pi r)Q_r = P$$

$$\text{i.e. } Q_r = \frac{-P}{2\pi r}$$

Fig. 5.6 Shear force due to
a concentrated load

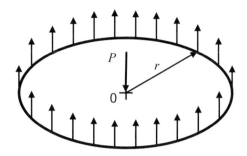

Hence,

$$\frac{d}{dr}\left\{\frac{1}{r}\frac{d}{dr}\left(r\frac{dw}{dr}\right)\right\} = \frac{P}{2\pi\, r D}$$

Successive integration and rearrangement yield

$$w = \frac{P}{8\pi\, D}r^2 \ln\frac{r}{a} + C_1 r^2 + C_2 \ln\frac{r}{a} + C_3$$

The deflection at the centre is finite, and hence $C_2 = 0$. Note that $Lt_{r\to 0}\left(r^2 \ln\frac{r}{a}\right) = 0$ by L'Hospital's rule.

The conditions at the outer periphery yield C_1 and C_3. Assuming that the plate is simply supported, the final solution is given by

$$w = \frac{P}{16\pi\, D}\left[2r^2 \ln\frac{r}{a} + \left(\frac{3+\mu}{1+\mu}\right)(a^2 - r^2)\right]$$

$$\text{and } w_{\max} = w_{\text{centre}} = \left(\frac{3+\mu}{1+\mu}\right)\frac{Pa^2}{16\pi\, D}$$

If the load, instead of being concentrated, is uniformly distributed over the plate, then from Example 5.2,

$$w_{\max} = \left(\frac{5+\mu}{1+\mu}\right)\left(\frac{P}{\pi\, a^2}\right)\frac{a^4}{64D} = \left(\frac{5+\mu}{1+\mu}\right)\frac{Pa^2}{64\pi\, D}$$

Thus the deflection increases by a factor of $\frac{4(3+\mu)}{(5+\mu)}$ when a uniformly distributed load is replaced by a statically equivalent central concentrated load. For most metals, $\mu \approx 0.3$, and this factor is 2.49.

Fig. 5.7 Axisymmetric
moments on an annular plate

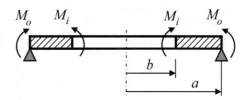

Example 5.4 Annular plate under axisymmetric moment loading.

The plate is simply supported at the outer periphery. Axisymmetric moments, of intensities M_i and M_o per unit length, are applied at the inner and the outer boundaries, respectively, as shown in Fig. 5.7.

For this case, $q = 0$, and hence,

$$w = C_1 + C_2 r^2 + C_3 \ln \frac{r}{a} + C_4 r^2 \ln \frac{r}{a}$$

The boundary conditions are:

$$w = 0, \quad M_r = M_o \text{ at } r = a$$

$$Q_r = 0, \quad M_r = M_i \text{ at } r = b$$

These conditions yield

$$C_1 = \frac{a^2(a^2 M_o - b^2 M_i)}{2D(1+\mu)(a^2 - b^2)} \quad C_2 = \frac{-(a^2 M_o - b^2 M_i)}{2D(1+\mu)(a^2 - b^2)}$$

$$C_3 = \frac{-a^2 b^2 (M_o - M_i)}{D(1-\mu)(a^2 - b^2)} \quad C_4 = 0$$

thus completing the solution.

One can specialize the above solution for a solid plate loaded by moment M_o at the outer boundary. This is done by putting $M_i = b = 0$ to obtain

$$C_1 = \frac{a^2 M_o}{2D(1+\mu)} \quad C_2 = \frac{-M_o}{2D(1+\mu)}$$

$$C_3 = C_4 = 0$$

and

$$w = \frac{M_o}{2D(1+\mu)}(a^2 - r^2)$$

Fig. 5.8 Moments along a
concentric circle

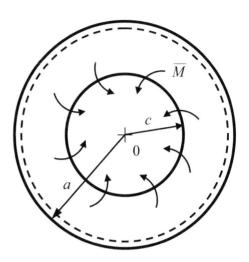

It should be noted that no reaction is offered by the simple support and so the
above result holds good for a free boundary as well, provided the deflections are
measured with reference to the boundary.

Example 5.5 Moments applied along a concentric circle on a simply supported plate.

The applied load is a moment of intensity $\overline{M}$ per unit length acting along the
circle of radius c as shown in Fig. 5.8. Looking at the free bodies corresponding to
the inner portion ($r \leq c$) and the outer annular portion separately, and using the fact
that there is no lateral load on the plate, it is clear that each portion is subjected to
moment loading as shown in Fig. 5.9, where M is yet to be determined. This can be
done by enforcing the continuity of the slope at $r = c$.

Fig. 5.9 Free body diagrams

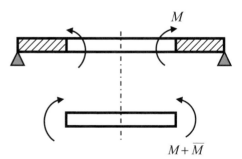

Using the results of Example 5.4, we have

for the outer portion:

$$w_{\text{outer}} = C_1 + C_2 r^2 + C_3 \ln \frac{r}{a} + C_4 r^2 \ln \frac{r}{a}$$

where $C_1 = \frac{-a^2 c^2 M}{2D(1+\mu)(a^2-c^2)}$ $C_2 = \frac{c^2 M}{2D(1+\mu)(a^2-c^2)}$

$C_3 = \frac{a^2 c^2 M}{D(1-\mu)(a^2-c^2)}$ $C_4 = 0$

for the inner portion:

$$w_{\text{inner}} = w_{\text{outer}}|_{r=c} + \frac{(M+\overline{M})(c^2 - r^2)}{2D(1+\mu)}$$

Equating the slopes $w_{\text{outer},r}$ and $w_{\text{inner},r}$ at $r = c$, one obtains

$$M = -\overline{M}(1-\mu)\frac{(a^2 - c^2)}{2a^2}$$

The final solution for the deflection is given by

$$w = -\frac{\overline{M}c^2}{2D}\ln\frac{r}{a} + \frac{\overline{M}c^2}{4D}\left(\frac{1-\mu}{1+\mu}\right)\frac{(a^2 - r^2)}{a^2} \quad \text{for } r \geq c$$

$$w = -\frac{\overline{M}c^2}{2D}\ln\frac{c}{a} + \frac{\overline{M}c^2}{2D(1+\mu)}$$

$$-\frac{\overline{M}}{4D(1+\mu)}\frac{r^2}{a^2}[a^2(1+\mu) + c^2(1-\mu)] \quad \text{for } r \leq c$$

Example 5.6

Ring load on a simply supported solid circular plate.

This problem (Fig. 5.10) has to be solved by realizing that the inner portion of the plate ($0 \leq r \leq c$) is free from shear force Q_r. Thus, the two portions correspond to the free bodies shown in Fig. 5.11, where M is an unknown moment. As in Example 5.5, M is found out by enforcing the continuity of the slope at $r = c$.

For the inner portion:

We have, from Example 5.4,

$$w_{\text{inner}} = \frac{M}{2D(1+\mu)}(c^2 - r^2) + w_{\text{outer}}\Big|_{r=c}$$

Fig. 5.10 Load along a
concentric circle

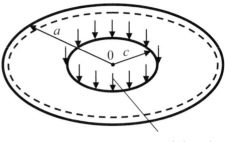

P per unit length

Fig. 5.11 Free body
diagrams

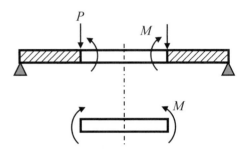

and hence

$$w_{\text{inner},r}\big|_{r=c} = \frac{-Mc}{D(1+\mu)}$$

For the outer portion:

$$w_{\text{outer}} = C_1 + C_2 r^2 + C_3 \ln\frac{r}{a} + C_4 r^2 \ln\frac{r}{a}$$

with C_1 to C_4 to be determined from the conditions:

$$w = M_r = 0 \text{ at } r = a$$

$$M_r = M, \quad Q_r = -P \text{ at } r = c$$

Carrying out this algebra, the slope of the outer portion at $r = c$ is obtained as

$$w_{\text{outer},r} = \frac{2Mc[a^2(1+\mu) + c^2(1-\mu)] - Pc^2\left[(a^2 - c^2)(1-\mu) + 2a^2(1+\mu)\ln\frac{a}{c}\right]}{2D(a^2 - c^2)(1-\mu^2)}$$

Equating the slopes for the two portions at $r = c$, one gets M in terms of P. The final solution for deflection turns out to be

$$w_{outer} = \frac{Pc}{8D(1+\mu)}\left(1 - \frac{r^2}{a^2}\right)[a^2(3-\mu) - c^2(1-\mu)] + \frac{Pc}{4D}(c^2+r^2)\ln\frac{r}{a}$$

$$w_{inner} = \frac{Pc}{8D(1+\mu)}\left(1 - \frac{c^2}{a^2}\right)[a^2(3+\mu) - r^2(1-\mu)] - \frac{Pc}{4D}(c^2+r^2)\ln\frac{a}{c}$$

The maximum deflection occurs at the centre and is given by

$$w_{max} = \frac{Pc}{8D}\left(\frac{3+\mu}{1+\mu}\right)(a^2 - c^2) - \frac{Pc^3}{4D}\ln\frac{a}{c}$$

Summary

The equations of the theory of elasticity and those of CPT have been presented with reference to cylindrical polar coordinates. For the special case of axisymmetric loading and axisymmetric boundary conditions, solutions have been obtained for a number of problems involving both solid and annular plates.

Conceptual Question

1. Suppose one is required to design a uniformly loaded, simply supported solid circular plate on the basis of the maximum shear stress. Give detailed step-wise instructions to do this. Be careful to consider both lateral surface and mid-surface locations. (*See Question 2 under* **Conceptual Questions** *of Chap. 4.*)

Problems

Sections 5.1 and 5.2

1. With reference to cylindrical polar coordinates, derive the equilibrium equations of CPT by direct integration of those of the theory of elasticity.
2. Derive the expression for the strain energy corresponding to axisymmetric deformation of a circular plate in terms of the deflection function.

Section 5.3

3. Find the solution for the deflection of a clamped circular plate under uniformly distributed load (a)directly; (b)by superposition of results available in this chapter. Plot the radial variation of w and M_r. Take $\mu = 0.3$.

4. A simply supported circular plate is subjected to axisymmetric distributed load varying linearly with respect to radius from zero at the centre to q_o at the outer periphery. Find the solution for the deflection.

5. Repeat Problem 4 if the load varies linearly from zero at the outer periphery to q_o at the centre.

6. An annular plate (0.15 m $\leq r \leq 0.5$ m) is simply supported at the outer periphery. Find the maximum deflection due to a uniform load of 30 N/m^2, if the inner periphery is (a) free; (b) guided such that the slope is completely restrained while deflection is freely permitted. Take $E = 200$ GPa, $\mu = 0.3$, and the thickness to be 2 mm.

7. Obtain an expression for the deflection of a clamped circular plate under a central concentrated load.

8. An annular plate with an inner diameter of 0.4 m and an outer diameter of 0.8 m is fixed at the inner boundary and free at the outer boundary. It is subjected to an axisymmetric distributed load with the intensity varying linearly from 3 kN/m^2 at the inner edge to 5 kN/m^2 at the outer edge. If the allowable bending stress is 75 MPa, determine the thickness required. Take $E = 210$ GPa and $\mu = 1/3$.

9. A simply supported circular plate of radius 0.4 m is loaded over a central concentric circular area of radius 0.1 m. The load is of uniform intensity of 1kN/m^2. Taking $E = 70$ GPa, $\mu = 0.3$, and the thickness to be 4 mm, find the maximum deflection and the maximum bending stress.

10. Using the result of Example 5.6, determine the flexural rigidity required for the plate shown if the central deflection is not to exceed 4 mm. Take $\mu = 0.3$. (*Hint: Find the ring load that produces the same deflection at the centre as a concentrated load at a given radius.*)

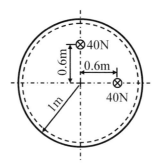

Problem 10

11. Explain how you would solve the problem corresponding to a circular patch
     load on a solid plate as shown. The actual solution is not expected.

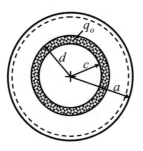

Problem 11

Chapter 6
Free and Forced Vibrations

In all the studies carried out in the earlier chapters, the loading was assumed to be static, i.e. not varying with time. However, in reality, structures are often subjected to time-varying loads or support motions—as, for example, due to a blast or an impulsive force, aerodynamic loads, inertia forces during accelera-tion or deceleration, seismic disturbances, vibrating machinery, and vibration during transport. In all such applications, the structure has to be designed such that the possibility of resonant vibration is minimized and the dynamic stresses are below the allowable stresses for the material. Towards this, dynamic anal-ysis has to be carried out both for identifying the resonant frequencies (*free vibration analysis*) and for finding out the dynamic displacements and stresses (*response analysis*). Such a study is described in this chapter with reference to simple problems involving rectangular and circular plates.

6.1 Equations of Motion

The equilibrium equations, derived earlier for static transverse loading, can be directly extended for the case of dynamic transverse loading by invoking D'Alembert's principle. To recollect, D'Alembert's principle can be stated as:

Every state of motion can be considered as a state of equilibrium under the action of inertia forces along with the external loads.

Thus, with reference to Sect. 2.3, the transverse force equilibrium equation [Eq. (2.14)] should now account for translational inertia while the two moment equi-librium equations [Eqs. (2.15) and (2.16)] should account for the inertia moment corresponding to rotation of the normal about x, y axes, respectively. However, it can be proved that the rotary inertia terms are negligible for a thin plate and hence only the translational inertia term need be considered in CPT. Thus, the only change required

© The Author(s) 2021
K. Bhaskar and T. K. Varadan, *Plates*,
https://doi.org/10.1007/978-3-030-69424-1_6

is the addition of the transverse inertia force per unit area to the transverse load q, and this change can be carried out directly in the governing biharmonic equation.

For a plate with mass per unit area $\bar{\rho}$, the inertia force is given by $-\bar{\rho}w_{,tt}$ per unit area. With this term, the governing equation for the dynamic problem can now be written as

$$D\nabla^4 w = q - \bar{\rho}w_{,tt}$$

$$\text{i.e.} \quad D\nabla^4 w + \bar{\rho}w_{,tt} = q \tag{6.1}$$

where $w = w(x, y, t)$ and $q = q(x,y, t)$. This equation can be used to solve for w once the time-varying transverse load q is specified and is hence the equation for response analysis.

For the case of free vibrations, q is zero, and the equation reduces to

$$D\nabla^4 w + \bar{\rho}w_{,tt} = 0 \tag{6.2}$$

Noting that a normal mode of vibration is one in which all elemental masses of the plate execute simple harmonic motion with the same frequency, the corresponding solution is sought as

$$w = W \sin \omega t \tag{6.3}$$

where W is a function of the space coordinates x,y. Thus, the problem is reduced to the equation

$$D\nabla^4 W - \bar{\rho}\omega^2 W = 0 \tag{6.4}$$

along with appropriate edge conditions.

Solutions for W, other than the trivial solution $W = 0$, are possible only for specific discrete values of the frequency ω; these eigenvalues are the desired natural frequencies of the plate, and the corresponding eigenfunctions W are the associated mode shapes.

For the case of a plate on elastic foundation, one has

$$D\nabla^4 w + k_F w + \bar{\rho}w_{,tt} = 0 \tag{6.5a}$$

leading to

$$D\nabla^4 W - (\bar{\rho}\omega^2 - k_F)W = 0 \tag{6.5b}$$

which is of the same form as Eq. (6.4); by virtue of this, all frequency results for this case can be obtained from the corresponding results without the foundation, as

$$\omega^2\big|_{\text{with foundation}} = \omega^2\big|_{\text{without foundation}} + \frac{k_F}{\rho} \tag{6.6}$$

6.2 Free Vibration Analysis

Simple free vibration solutions for rectangular plates with some specific edge conditions are possible by Navier and Levy methods. These are similar to the solutions for static flexure presented earlier and are explained below, followed by a slightly more complicated solution involving Bessel functions for a circular plate.

Example 6.1 Free vibrationsof a simply supported rectangular plate (Navier's method).

The edge conditions for the plate ($0\ x \leq a; 0 \leq y \leq b$) are

$$W = W_{,xx} = 0 \text{ at } x = 0, a$$

$$W = W_{,yy} = 0 \text{ at } y = 0, b$$

Seeking, by Navier's method, a solution for W in the form

$$W = \sum_m \sum_n W_{mn} \sin\frac{m\pi x}{a} \sin\frac{n\pi y}{b}$$

which satisfies the edge conditions, one gets

$$\sum_m \sum_n \left[D\left\{ \left(\frac{m\pi}{a}\right)^4 + 2\left(\frac{m\pi}{a}\right)^2\left(\frac{n\pi}{b}\right)^2 + \left(\frac{n\pi}{b}\right)^4 \right\} - \bar{\rho}\omega^2 \right]$$
$$W_{mn} \sin\frac{m\pi x}{a} \sin\frac{n\pi y}{b} = 0$$

i.e.

$$\left[D\left\{ \left(\frac{m\pi}{a}\right)^4 + 2\left(\frac{m\pi}{a}\right)^2\left(\frac{n\pi}{b}\right)^2 + \left(\frac{n\pi}{b}\right)^4 \right\} - \bar{\rho}\omega^2 \right] W_{mn} = 0 \quad \text{for all } m, n$$

Since W_{mn} is non-zero for non-trivial solutions, we have, for any m and n,

$$D\left\{ \left(\frac{m\pi}{a}\right)^4 + 2\left(\frac{m\pi}{a}\right)^2\left(\frac{n\pi}{b}\right)^2 + \left(\frac{n\pi}{b}\right)^4 \right\} - \bar{\rho}\omega_{mn}^2 = 0$$

which is the frequency equation.

Thus, the natural frequencies are

$$\omega_{mn} = \sqrt{\frac{D}{\rho}\left[\left(\frac{m\pi}{a}\right)^2 + \left(\frac{n\pi}{b}\right)^2\right]} \quad \text{for } m, n = 1, 2, \text{etc.}$$

corresponding to the mode shapes

$$W(x, y) = \sin\frac{m\pi x}{a}\sin\frac{n\pi y}{b}.$$

The fundamental frequency corresponds to $m = n = 1$ and is given by

$$\omega_{11} = \pi^2\sqrt{\frac{D}{\rho}\left(\frac{1}{a^2} + \frac{1}{b^2}\right)}$$

Considering a specific plate with $a = 2b$ and listing the frequencies $\left(\frac{\bar{\rho}\omega^2 b^4}{\pi^4 D}\right)$ in increasing order, one has the values

1.56 (1, 1), 4 (2, 1), 10.6 (3, 1), 18.1 (1, 2),
25 (4,1) & (2, 2), 39.1 (3, 2), 52.6 (5, 1), 64 (4, 2), etc.

with m, n given in parentheses.

From these results, one should note that the frequencies increase as the number of half-waves in any direction increases, but modes with $m > n$ occur before those with $n > m$. This is because it is easier for the plate to bend in the longer direction.

Example 6.2 Free vibration analysis of a rectangular plate with two opposite edges simply supported (Levy's method).

The plate occupies the domain $(0 \times \leq a, -b/2 \leq y \leq b/2)$ and is simply supported at $x = 0$ and $x = a$. Levy's method is applicable for arbitrary support conditions at $y = \pm b/2$, and these edges are taken to be clamped in the present example.

Seeking a solution in the form

$$W = \sum_m W_m(y)\sin\frac{m\pi x}{a}$$

which satisfies the simple-support conditions,

$$W = W_{,xx} = 0 \quad \text{at } x = 0, a,$$

one obtains

$$\sum_m \left[D\left\{ W_{m,yyyy} - 2\left(\frac{m\pi}{a}\right)^2 W_{m,yy} + \left(\frac{m\pi}{a}\right)^4 W_m \right\} - \overline{\rho}\omega^2 W_m \right] \sin \frac{m\pi x}{a} = 0$$

i.e. $W_{m,yyyy} - 2\left(\frac{m\pi}{a}\right)^2 W_{m,yy} + \left(\frac{m\pi}{a}\right)^4 W_m - \frac{\overline{\rho}\omega^2 W_m}{D} = 0$ for all m.

This equation has the general solution given by

$$W_m = A_m \cosh \alpha y + B_m \sinh \alpha y + C_m \cos \beta y + D_m \sin \beta y$$

where

$$\alpha = \sqrt{\left[\sqrt{\frac{\overline{\rho}\omega^2}{D}} + \left(\frac{m\pi}{a}\right)^2 \right]} \quad \text{and} \quad \beta = \sqrt{\left[\sqrt{\frac{\overline{\rho}\omega^2}{D}} - \left(\frac{m\pi}{a}\right)^2 \right]}$$

and A_m to D_m are constants to be determined from the clamped edge conditions

$$W_m = W_{m,y} = 0 \quad \text{at } y = \pm b/2.$$

These lead to

$$A_m \cosh \tfrac{\alpha b}{2} + B_m \sinh \tfrac{\alpha b}{2} + C_m \cos \tfrac{\beta b}{2} + D_m \sin \tfrac{\beta b}{2} = 0$$
$$A_m \cosh \tfrac{\alpha b}{2} - B_m \sinh \tfrac{\alpha b}{2} + C_m \cos \tfrac{\beta b}{2} - D_m \sin \tfrac{\beta b}{2} = 0$$
$$\alpha A_m \sinh \tfrac{\alpha b}{2} + \alpha B_m \cosh \tfrac{\alpha b}{2} - \beta C_m \sin \tfrac{\beta b}{2} + \beta D_m \cos \tfrac{\beta b}{2} = 0$$
$$-\alpha A_m \sinh \tfrac{\alpha b}{2} + \alpha B_m \cosh \tfrac{\alpha b}{2} + \beta C_m \sin \tfrac{\beta b}{2} + \beta D_m \cos \tfrac{\beta b}{2} = 0$$

which can be rewritten as

$$A_m \cosh \frac{\alpha b}{2} + C_m \cos \frac{\beta b}{2} = 0$$

$$\alpha A_m \sinh \frac{\alpha b}{2} - \beta C_m \sin \frac{\beta b}{2} = 0$$

$$B_m \sinh \frac{\alpha b}{2} + D_m \sin \frac{\beta b}{2} = 0$$

$$\alpha B_m \cosh \frac{\alpha b}{2} + \beta D_m \cos \frac{\beta b}{2} = 0$$

The decoupling of the constants A_m and C_m with respect to the constants B_m and D_m indicates that there are two sets of frequencies, the first corresponding to mode shapes even in y and the second to those odd in y. These are referred to as y-symmetric and y-antisymmetric mode shapes, respectively.

Seeking non-trivial solutions, the frequency equation corresponding to y-symmetric mode shapes is given by

$$\begin{vmatrix} \cosh\frac{\alpha b}{2} & \cos\frac{\beta b}{2} \\ \alpha\sinh\frac{\alpha b}{2} & -\beta\sin\frac{\beta b}{2} \end{vmatrix} = 0$$

and that for y-antisymmetric mode shapes is

$$\begin{vmatrix} \sinh\frac{\alpha b}{2} & \sin\frac{\beta b}{2} \\ \alpha\cosh\frac{\alpha b}{2} & \beta\cos\frac{\beta b}{2} \end{vmatrix} = 0$$

Since $\alpha^2 = \beta^2 + 2\left(\frac{m\pi}{a}\right)^2$, the above frequency equations are transcendental equations in β, each having an infinite number of roots for every m. The equations can be solved by Newton–Raphson method or graphically. Corresponding to any root (i.e. β and hence ω), one gets the mode shape in terms of either (A_m/C_m) with $B_m = D_m = 0$, or (B_m/D_m) with $A_m = C_m = 0$.

At this stage, let us consider a square plate $(b = a)$, and find the frequencies corresponding to the first y-symmetric and y-antisymmetric modes. Taking $m = 1$, and putting $\alpha a = \sqrt{[(\beta a)^2 + 2\pi^2]}$, the frequency equation for the y-symmetric modes can be written as

$$-\beta a\sin\frac{\beta a}{2}\cosh\frac{\sqrt{[(\beta a)^2 + 2\pi^2]}}{2}$$
$$-\sqrt{[(\beta a)^2 + 2\pi^2]}\cos\frac{\beta a}{2}\sinh\frac{\sqrt{[(\beta a)^2 + 2\pi^2]}}{2} = 0$$

which has the roots $\beta a = 4.3682,\ 10.9191,\ 17.2466$, etc. The corresponding non-dimensional frequency parameter $\frac{\bar{\rho}\omega^2 a^4}{D}$ is equal to 838.15, 16,665.7, 94,443.2, etc. The mode shapes obtained are given by putting $(A_1/C_1) = 0.05097,\ -0.003745,\ 0.0001888$, etc., with $B_1 = D_1 = 0$ and $m = 1$ in the expression for W. These mode shapes correspond to 0, 2, 4, ... nodal lines parallel to the x-axis.

For the y-antisymmetric modes with $m = 1$, proceeding in a similar fashion, one obtains $\beta a = 7.7109,\ 14.0898,\ 20.3972$, etc., $\frac{\bar{\rho}\omega^2 a^4}{D} = 4806.23,\ 43,427.1,\ 181,403.1$, etc., and $(B_1/D_1) = 0.01530,\ -0.0008549,\ 0.00004096$, etc. with $A_1 = C_1 = 0$. These values correspond to 1, 3, 5, ... nodal lines parallel to the x-axis.

It should be clear from the above results that the frequencies for the y-symmetric modes and those for the y-antisymmetric modes are distinct from each other, and that they increase in magnitude as the number of nodal lines parallel to the x-axis increases. By carrying out calculations for $m = 2, 3$, etc., it can be shown that for a fixed number of nodal lines parallel to the x-axis, the frequency increases as m increases. Thus, the fundamental frequency for the plate under consideration corresponds to $m = 1$ and no nodal lines parallel to the x-axis; its value is given by $\frac{\bar{\rho}\omega^2 a^4}{D} = 838.15$.

Example 6.3 Free vibrations of a solid circular plate.

The biharmonic operator with reference to polar coordinates was given earlier in Eq. (5.11). In view of the axisymmetry of the geometry and boundary conditions, a solution can be sought in the form

$$W(r, \theta) = \sum_{n=0,1,\ldots} W_n(r) \cos n\theta$$

Though this seems to be a solution only for modes symmetric about $\theta = 0$, the antisymmetric modes are also included therein in view of the axisymmetry of the plate; for example, the mode corresponding to $n = 1$ can be looked upon as an antisymmetric mode about $\theta = 90°$. The nth term represents a mode with n nodal diameters, while the first term ($n = 0$) corresponds to the axisymmetric mode without any nodal diameters.

Substitution of this solution in Eq. (6.4) reduces it to

$$\left(\frac{d^2}{dr^2} + \frac{1}{r}\frac{d}{dr} - \frac{n^2}{r^2} \right)\left(\frac{d^2}{dr^2} + \frac{1}{r}\frac{d}{dr} - \frac{n^2}{r^2} \right)W - \frac{\bar{\rho}\omega^2}{D}W = 0$$

which can be rewritten as

$$\left(\frac{d^2}{dr^2} + \frac{1}{r}\frac{d}{dr} - \frac{n^2}{r^2} + \lambda^2 \right)\left(\frac{d^2}{dr^2} + \frac{1}{r}\frac{d}{dr} - \frac{n^2}{r^2} - \lambda^2 \right)W = 0$$

where $\lambda^4 = \frac{\bar{\rho}\omega^2}{D}$.

The four linearly independent solutions of the above equation can be sought as two each of the following second-order equations

$$\left(\frac{d^2}{dr^2} + \frac{1}{r}\frac{d}{dr} + \left(\lambda^2 - \frac{n^2}{r^2}\right) \right)W = 0$$

$$\left(\frac{d^2}{dr^2} + \frac{1}{r}\frac{d}{dr} - \left(\lambda^2 + \frac{n^2}{r^2}\right) \right)W = 0$$

These can be recognized as Bessel's equation and modified Bessel's equation, respectively. Putting the corresponding solutions together, the general solution for W can be expressed as

$$W_n = A_n J_n(\lambda r) + B_n Y_n(\lambda r) + C_n I_n(\lambda r) + D_n K_n(\lambda r)$$

where J_n and Y_n are Bessel functions of the first and second kinds, and I_n and K_n their modified counterparts.

For a solid circular plate, the functions Y_n and K_n lead to infinitely large deflections and stresses at the centre $r = 0$ and hence the corresponding constants B_n and D_n have to be taken as zero. The other two constants have to be determined from the outer edge conditions.

Considering clamped conditions at the outer edge.

i.e. $W_n = W_{n,r} = 0$ at $r = a$

and seeking non-trivial solutions for A_n and C_n, one gets the frequency equation as

$$J_n(\lambda a) I'_n(\lambda a) - I_n(\lambda a) J'_n(\lambda a) = 0$$

which can be solved graphically or by trial and error to obtain λ's and hence the natural frequencies. Corresponding to each λ, the ratio A_n/C_n can be obtained and hence the shape of $W_n(r)$.

As is intuitively expected, the mode shapes will correspond to various numbers of nodal diameters and nodal circles with the fundamental mode having neither of them. For the clamped plate, the next two modes correspond to shapes with 1 and 2 nodal diameters while the fourth mode shape is axisymmetric with one nodal circle. The actual values of the frequencies and radii of the nodal circles for several common boundary conditions can be found in handbooks.

6.3   Forced Vibration Analysis

We now consider the forced vibration of a plate due to transverse load varying harmonically with time. Considering the steady state response alone, the deflection at any point of the plate also varies harmonically with the same frequency as the excitation. If the mode shapes of the plate are known, then the unknown dynamic deflection can be sought as a summation of the mode shapes. This is illustrated below with reference to a simply supported plate.

Example 6.4 Harmonic excitation of a simply supported square plate.

Consider a simply supported square plate ($0 \leq x, y \leq a$) subjected to uniformly distributed load varying harmonically with time as

$$q(x, y, t) = q_o \sin \Omega t$$

The natural frequencies and mode shapes of this plate are given by

$$\omega_{mn} = \frac{\pi^2}{a^2} \sqrt{\frac{D}{\rho}} (m^2 + n^2) \quad \text{and} \quad \sin \frac{m\pi x}{a} \sin \frac{n\pi y}{a}$$

Noting that the deflection also varies harmonically with time as

$$w = W(x, y) \sin \Omega t$$

the equation of motion is reduced to

$$DV^4 W - \overline{\rho}\Omega^2 W = q_o$$

Seeking a solution for W as a summation of the normal modes

$$W = \sum_m \sum_n W_{mn} \sin \frac{m\pi x}{a} \sin \frac{n\pi y}{a}$$

so that the boundary conditions are satisfied a priori, and expanding the applied load also in a similar series as

$$q_o = \sum_{m=1,3,...} \sum_{n=1,3,...} \frac{16q_o}{mn\pi^2} \sin \frac{m\pi x}{a} \sin \frac{n\pi y}{a}$$

one gets, for odd m,n,

$$W_{mn} = \frac{16q_o}{mn\pi^2 \left[\frac{D\pi^4}{a^4}\left(m^2 + n^2\right)^2 - \overline{\rho}\Omega^2 \right]} = \frac{16q_o}{mn\pi^2 \overline{\rho}(\omega_{mn}^2 - \Omega^2)}$$

which coincides with the Navier solution for static loading (see Example 4.1) when $\Omega = 0$.

The coefficient W_{mn}, for any specific m,n, gives the contribution of that particular mode shape to the dynamic response and is hence referred to as the corresponding modal participation factor. When the exciting frequency coincides with any of the natural frequencies, the corresponding modal participation factor goes to infinity indicating a resonant response in that particular mode. In view of a small amount of damping that is invariably present, the resonant response would be large but finite, and the deformed shape of the plate would be that corresponding to the particular mode shape; this is true irrespective of whether the applied transverse load is uniform or some other arbitrary distribution. This has to be contrasted with the case of static loading where the deformed shape depends on the actual load distribution.

When the exciting frequency is close to one of the natural frequencies, the corresponding modal participation factor will be significantly higher compared to those for the other modes, and the plate deformation would be more or less the same as that particular mode shape. Thus, even for uniform loading, the dynamic deflection may have a wavy distribution—for example, three half-waves in each direction for $\Omega \approx \omega_{33}$.

A plot of the central deflection versus the exciting frequency is shown in Fig. 6.1. The central deflection is normalized with respect to that for static loading and the exciting frequency with respect to the first natural frequency ω_{11}. The peaks occurring at $\Omega/\omega_{11} = 5, 9$ correspond to resonances at ω_{13} and ω_{33}, respectively. The negative deflections indicate a completely out-of-phase response with respect to the excitation.

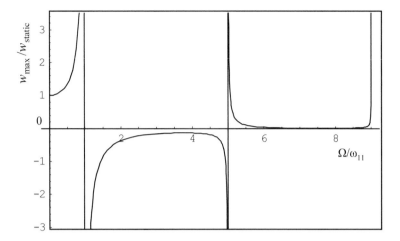

Fig. 6.1 Magnification factor for central deflection versus Ω/ω_{11}

Summary

Some simple free and forced vibration solutions have been illustrated in this chapter. It should be noted that only a few problems can be solved rigorously as explained above while most practical plate vibration problems require the use of approximate analytical and numerical methods.

Conceptual Question

1. Consider square plates with the following boundary conditions: (a) SCSC; (b) SFSF; (c) SSSS; (d) CCCC; (e) SSSC; (f) SSSF. Assuming the geometry and the material to be same for all the plates, list them in the order of increasing fundamental frequency. Explain why.

Problems

1. What is the percentage change in the fundamental frequency of a square plate, simply supported all around, when

   (a) the side of the square is doubled?
   (b) the thickness is doubled?
   (c) the material is changed from aluminium to steel?

2. In Example 6.2, results for the frequencies and the corresponding mode shapes of a square plate have been given. Starting from the frequency equation, verify that the results presented for the first y-symmetric and y-antisymmetric modes with $m = 1$ are correct. Plot the two mode shapes with respect to y.

3. Using Levy's method, obtain the frequency equation for a plate with $a/b = 2$ and with all edges simply supported. Using the result of Example 6.1, calculate the first three frequencies and verify that they satisfy the frequency equation obtained by Levy's method. Show that the mode shapes yielded by Levy's method are the same as those of Example 6.1.

4. Using Levy's method, find the frequency equation of a square plate simply supported on three sides and clamped on the fourth (at $y = a/2$). Show that $\omega a^2 \sqrt{\frac{\rho}{D}} = 23.6464$ corresponds to a solution for $m = 1$; find the associated mode shape and plot it along y.

Chapter 7
Effect of In-Plane Forces on Static Flexure, Dynamics and Stability

When a plate is subjected to in-plane forces, such forces have a stiffening or slackening effect on flexural behaviour and hence on the static deflections due to any applied transverse loading and also the flexural vibration frequencies. Further, a plate subjected to in-plane forces alone may become unstable in the flat configuration and undergo buckling. Design for stability is very important in current engineering practice because the availability of high-strength materials has led to the use of very thin plate structures. These aspects are discussed in this chapter.

7.1 Governing Equations for Combined Bending and Stretching

While deriving the equilibrium equations for the problem of flexure due to transverse loads alone (Sect. 2.3), the undeformed plate configuration was taken as the reference. To account for the effect of in-plane forces on flexural behaviour, one has to consider the equilibrium of the bent configuration. It should be noted that this will be done within the purview of the linear small deformation theory and with the assumption that the intensities of the in-plane forces at any point of the plate do not vary as the plate bends.

Consider a small element of the plate, of dimensions dx x dy, as shown in Fig. 7.1, where attention is focussed on the effect of the internal in-plane forces, namely, the normal forces N_x and N_y per unit length and the shear forces N_{xy} and N_{yx} per unit length. As was done in Sect. 2.3, the incremented forces are approximated by a truncated Taylor expansion; these are denoted using single and double stars as

© The Author(s) 2021
K. Bhaskar and T. K. Varadan, *Plates*,
https://doi.org/10.1007/978-3-030-69424-1_7

Fig. 7.1 In-plane forces on a deformed plate element

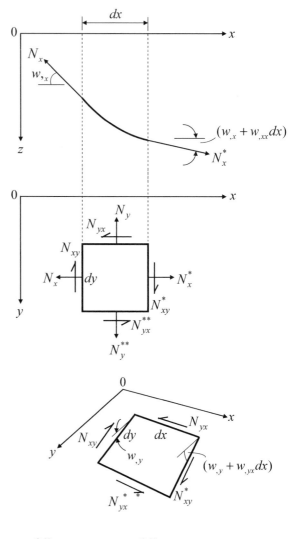

$$() ^* = () + \frac{\partial ()}{\partial x}dx \ () ^{**} = () + \frac{\partial ()}{\partial y}dy$$

All the forces shown in Fig. 7.1 are positive as per the sign convention adopted here. The plate element is in equilibrium under the action of these in-plane forces along with the transverse forces and moments (q, Q_x, Q_y, M_x, M_y and M_{xy} and their incremented counterparts) as shown earlier in Fig. 2.3. However, unlike in Fig. 2.3, all the forces acting on the deformed element here have small inclinations with respect to the corresponding coordinate axes, as is clearly shown in Fig. 7.1 for N_x and N_{xy}. These inclinations are also incremented using a truncated Taylor expansion in the x and y directions.

For force equilibrium in the x-direction

$$(N_x + N_{x,x}dx)dy \cos(w_{,x} + w_{,xx}dx) - N_x dy \cos w_{,x}$$
$$+ [(N_{yx} + N_{yx,y}dy)dx - N_{yx}dx] \cos\left(w_{,x} + w_{,xx}\frac{dx}{2}\right) = 0$$

Putting $\cos\theta \approx 1$ for small deformations, one obtains

$$N_{x,x} + N_{yx,y} = 0 \tag{7.1}$$

While summing the above forces in the x-direction, the small contributions of Q_x, Q_x^*, Q_y, Q_y^*, which are all slightly inclined with respect to the z-direction, have been neglected.

Similarly, by summing forces in the y-direction, one gets

$$N_{xy,x} + N_{y,y} = 0 \tag{7.2}$$

Satisfaction of moment equilibrium about z-axis yields, after neglect of higher-order terms,

$$N_{xy} = N_{yx} \tag{7.3}$$

For force equilibrium in the z-direction, in addition to transverse load q and the shear forces at the edges, one has to include the z-components of the in-plane forces. These additional quantities are

(a) $(N_x + N_{x,x}dx)dy \sin(w_{,x} + w_{,xx}dx) - N_x dy \sin w_{,x}$
     which, for small deformations (i.e. $\sin\theta \approx \theta$), and after neglect of the term involving $dx^2 dy$, becomes $(N_x w_{,xx} + N_{x,x}w_{,x})dx\,dy$;
(b) $(N_y w_{,yy} + N_{y,y}w_{,y})dx\,dy$ obtained similar to (a) above;
(c) $(N_{xy} + N_{xy,x}dx)dy \sin(w_{,y} + w_{,xy}dx) - N_{xy}dy \sin w_{,y}$
     which becomes $(N_{xy}w_{,xy} + N_{xy,x}w_{,y})dx\,dy$;
(d) $(N_{yx}w_{,xy} + N_{yx,y}w_{,x})dx\,dy$ obtained similar to (c) above.

Thus, due to the net additional effect of the in-plane forces, the z-direction equilibrium equation gets modified from Eq. (2.14) to

$$Q_{x,x} + Q_{y,y} + q + N_x w_{,xx} + N_y w_{,yy} + 2N_{xy}w_{,xy}$$
$$+ N_{x,x}w_{,x} + N_{y,y}w_{,y} + N_{xy,x}w_{,y} + N_{xy,y}w_{,x} = 0$$

By virtue of Eqs. (7.1) and (7.2), the coefficients of $w_{,x}$ and $w_{,y}$ turn out to be zero, and hence the above equation reduces to

$$Q_{x,x} + Q_{y,y} + q + N_x w_{,xx} + N_y w_{,yy} + 2N_{xy}w_{,xy} = 0 \tag{7.4}$$

The moment equilibrium equations with respect to the x and y axes are not affected by the in-plane forces because the contributions due to these forces are of higher-order and hence negligible compared to the other quantities considered in Sect. 2.3. For the sake of completeness, these two equations are reproduced below.

$$M_{xy,x} + M_{y,y} - Q_y = 0 \tag{7.5}$$

$$M_{x,x} + M_{xy,y} - Q_x = 0 \tag{7.6}$$

Equations (7.1)–(7.3) pertain to the in-plane problem, i.e. the problem of determining the internal in-plane force field (N_x, N_y, N_{xy}). This problem can be solved once the in-plane tractions on the edges of the plate are specified; this is done without considering the transverse deformation at all because the in-plane force field is taken to be unaffected by it. Once the in-plane force field is obtained, its influence on the flexural behaviour of the plate is accounted for by solving Eqs. (7.4)–(7.6). Thus, the present small deformation formulation accounts for just one-way coupling between the in-plane forces and transverse deflections. (In contrast, for a problem involving finite deformations, bending is accompanied by in-plane stretching and the in-plane forces increase as the plate bends more and more, thus leading to a non-linear theory. This will be discussed in Chap. 14).

For a rectangular plate ($0 \le x \le a, 0 \le y \le b$) with

(a)   uniform N_x applied along $x = 0$ and $x = a$, or
(b)   uniform N_y applied along $y = 0$ and $y = b$, or
(c)   uniform N_{xy} applied along all the four edges or
(d)   any combination of (a)–(c),

the solution of Eqs. (7.1)–(7.3) is simply a uniform internal force field over the entire area; only such cases are considered in the present chapter. (Varying internal force fields due to non-uniform edge loading will be considered later in Chap. 13).

The bending problem can be reduced, as was done in Sect. 2.4, to a single governing equation in w by eliminating Q_x and Q_y from Eqs. (7.4)–(7.6) and by using the moment–curvature relations.

This final governing equation is given by

$$D\nabla^4 w = q + N_x w_{,xx} + N_y w_{,yy} + 2N_{xy} w_{,xy} \tag{7.7}$$

and is the mathematical statement of the equilibrium of the deflected plate due to any combination of transverse and in-plane forces.

When the in-plane forces are absent, the above equation reduces to the biharmonic equation derived earlier. When in-plane forces are present besides transverse loads, the equation accounts for the stiffening or slackening influence of the in-plane forces and yields correspondingly lower or higher deflections. When in-plane forces alone are present, a deflected equilibrium configuration is possible only if the plate has buckled, and hence the governing equation pertains to the buckling problem.

While the equilibrium configurations corresponding to the first two cases mentioned above (i.e. q alone, and q along with N_x, N_y and N_{xy}) are stable equilibrium configurations, the last case (N_x, N_y and N_{xy} alone) is one of neutral equilibrium. The governing equation is non-homogeneous as long as q is present yielding a unique solution (*equilibrium problem*), while it becomes homogeneous when q is zero, in which case w is indeterminate and is non-trivial only for certain specific values of the in-plane forces (*eigenvalue problem*).

7.2 Analysis for Stability

We shall start with the stability problem and illustrate the application of Navier and Levy methods to determine critical loads for certain rectangular plate problems.

Example 7.1 Buckling of a simply supported plate (Navier's method)

Let the plate be subjected to biaxial compression with uniform compressive force N_x along the edges $x = 0$ and $x = a$, and uniform compressive force N_y along the edges $y = 0$ and $y = b$. Let N_{xy} be absent. (If N_{xy} is present, the problem can be solved only by an approximate method—see Chap. 8). N_x and N_y are increased simultaneously, keeping their relative proportion constant, till buckling occurs. Let the critical values be N_{xcr} and N_{ycr} and let

$$N_{x\,cr} = -k_1 P_{cr}$$
$$N_{y\,cr} = -k_2 P_{cr}$$

where P_{cr} has the units of force per unit length and the negative sign indicates compression.

If w is taken as

$$w = \sum_m \sum_n W_{mn} \sin \frac{m\pi x}{a} \sin \frac{n\pi y}{b}$$

such that the support conditions are satisfied a priori, Eq. (7.7) becomes (with $q = 0$),

$$D \sum \sum \left[\left(\frac{m\pi}{a}\right)^4 + 2\left(\frac{m\pi}{a}\right)^2\left(\frac{n\pi}{b}\right)^2 + \left(\frac{n\pi}{b}\right)^4 \right] W_{mn} \sin \frac{m\pi x}{a} \sin \frac{n\pi y}{b}$$
$$= \sum \sum \left[-N_{x\,cr}\left(\frac{m\pi}{a}\right)^2 - N_{y\,cr}\left(\frac{n\pi}{b}\right)^2 \right] W_{mn} \sin \frac{m\pi x}{a} \sin \frac{n\pi y}{b}$$

i.e.

$$D\left[\left(\frac{m\pi}{a}\right)^2 + \left(\frac{n\pi}{b}\right)^2 \right]^2 + N_{x\,cr}\left(\frac{m\pi}{a}\right)^2 + N_{y\,cr}\left(\frac{n\pi}{b}\right)^2 = 0 \quad \text{for all } m, n$$

i.e.

$$P_{cr} = \frac{D\left[\left(\frac{m\pi}{a}\right)^2 + \left(\frac{n\pi}{b}\right)^2\right]^2}{\left[k_1\left(\frac{m\pi}{a}\right)^2 + k_2\left(\frac{n\pi}{b}\right)^2\right]} \quad \text{for all } m, n.$$

By taking different values of k_1 and k_2, one can specialize for the cases of uniaxial compression along x or y and for different cases of biaxial compression as explained below.

Case (a): Uniaxial compression in the x-direction ($k_1 = 1$, $k_2 = 0$)
Here,

$$N_{xcr} = -P_{cr} = \frac{-D\left[\left(\frac{m\pi}{a}\right)^2 + \left(\frac{n\pi}{b}\right)^2\right]^2}{\left(\frac{m\pi}{a}\right)^2}$$

$$= \frac{-\pi^2 a^2 D}{m^2}\left(\frac{m^2}{a^2} + \frac{n^2}{b^2}\right)^2 \quad \text{for all } m \text{ and } n.$$

Thus there are an infinite number of buckling loads, corresponding to various (m, n) combinations, associated with the corresponding normalized mode shapes given by

$$w(x, y) = \sin\frac{m\pi x}{a} \sin\frac{n\pi y}{b}$$

Of these, the lowest buckling load is of practical interest. To find this out, the (m,n) combination yielding the minimum P_{cr} has to be identified. In the expression for N_{xcr}, n appears only in the numerator, and hence $n = 1$ for the lowest buckling load. To fix up m, N_{xcr} is rewritten as

$$N_{x\,cr} = \frac{-\pi^2 D}{b^2}\left(\frac{mb}{a} + \frac{a}{mb}\right)^2$$

Thus, the required value of m is obtained by putting

$$\frac{d}{dm}\left[\left(\frac{mb}{a} + \frac{a}{mb}\right)^2\right] = 0$$

yielding m to be a/b.

Since m can take only integral values, the above result is applicable for integral values of a/b; in such cases, the plate buckles into square portions as shown in Fig. 7.2 with adjacent squares deflecting in opposite directions.

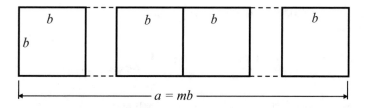

Fig. 7.2 Buckling into square portions

The corresponding buckling load is given by

$$N_{x\,\mathrm{cr}} = \frac{-4\pi^2 D}{b^2}$$

If a/b is not an integer, one has to check for both the integral numbers around a/b. For example, if $a/b = 1.2$, one has to check with $m = 1$ and $m = 2$ to identify the lowest buckling load. Instead, one can plot $\left(\frac{mb}{a} + \frac{a}{mb}\right)^2$ vs. a/b for different integer values of m, as shown in Fig. 7.3.

At $a/b = \sqrt{2}$, the value of $N_{x\mathrm{cr}}$ is same for both $m = 1$ and $m = 2$, indicating that the plate can buckle into either of these mode shapes ($m = n=1$ and $m = 2, n = 1$) with equal ease. Such transition points occur at $a/b = \sqrt{m(m + 1)}$ as shown. So, for any given non-integral a/b, depending on whether it falls to the left or right of a transition point, the critical value of m is easily identified.

Case (b): Uniaxial compression in the y-direction ($k_1 = 0$, $k_2 = 1$)

This is similar to Case (a). Here, corresponding to the lowest buckling load, m is equal to 1, while n depends on b/a.

Fig. 7.3 Identification of m

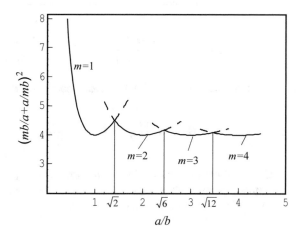

Case (c): Equal biaxial compression ($k_1 = k_2 = 1$)

Here, one gets

$$P_{cr} = D\left[\left(\frac{m\pi}{a}\right)^2 + \left(\frac{n\pi}{b}\right)^2\right] \quad \text{for each } (m, n)$$

The lowest value corresponds to $m = n = 1$, leading to

$$P_{cr} = D\pi^2\left(\frac{1}{a^2} + \frac{1}{b^2}\right) = \frac{\pi^2 D}{b^2}\left[1 + \left(\frac{b}{a}\right)^2\right]$$

For a square plate, $P_{cr} = \frac{2\pi^2 D}{b^2}$, and this is half that for uniaxial compression. For integral values of a/b, i.e. 2,3,4, etc., one can easily see that P_{cr} approaches $\frac{\pi^2 D}{b^2}$ as a/b increases, and this value is one-fourth that for uniaxial compression. In all these cases, P_{cr} is less than that for uniaxial compression because of the slackening influence of the additional compressive stress. In contrast, if one of the biaxial stresses is tensile, the buckling load will be more than that for uniaxial compression as illustrated below.

Case(d): Buckling due to N_{xcr} in the presence of constant tension $N_y = \overline{N}$

Here N_x alone is increased till buckling occurs, while N_y is kept constant. One has to start with

$$D\left[\left(\frac{m\pi}{a}\right)^2 + \left(\frac{n\pi}{b}\right)^2\right]^2 + N_{x\,cr}\left(\frac{m\pi}{a}\right)^2 + \overline{N}\left(\frac{n\pi}{b}\right)^2 = 0 \quad \text{for all } m, n$$

i.e.

$$N_{xcr} = \frac{-\left[D\left\{\left(\frac{m\pi}{a}\right)^2 + \left(\frac{n\pi}{b}\right)^2\right\}^2 + \overline{N}\left(\frac{n\pi}{b}\right)^2\right]}{\left(\frac{m\pi}{a}\right)^2}$$

$$= \frac{-\pi^2 a^2 D}{m^2}\left(\frac{m^2}{a^2} + \frac{n^2}{b^2}\right)^2 - \overline{N}\left(\frac{an}{bm}\right)^2 \quad \text{for all } m, n$$

It is clear that the lowest buckling load occurs for $n = 1$ and is given by

$$N_{x\,cr} = \frac{-\pi^2 D}{b^2}\left[\left(\frac{mb}{a} + \frac{a}{mb}\right)^2\right] - \overline{N}\left(\frac{a}{mb}\right)^2$$

The value of m corresponding to the lowest $N_{xnn\,ncr}$ depends on a/b and can be determined as done earlier—one should note that the buckling load will always be higher than the value corresponding to $\overline{N} = 0$.

Example 7.2 Buckling of a rectangular plate with two opposite edges simply supported (Levy's method)

Consider the case of a plate ($0 \leq x \leq a$, $-b/2 \leq y \leq b/2$) simply supported at $x = 0$ and $x = a$, clamped at $y = \pm b/2$, and buckling under uniaxial compression given by $N_{xcr} = -P_{cr}$.

The governing equation becomes

$$D\nabla^4 w = -P_{cr} w_{,xx}$$

Choosing the deflection as

$$w = \sum_m W_m(y) \sin \frac{m\pi x}{a}$$

so that the simply supported edge conditions are satisfied a priori, the governing equation reduces to

$$\sum_m \left[D \left\{ W_{m,yyyy} - 2 \left(\frac{m\pi}{a} \right)^2 W_{m,yy} + \left(\frac{m\pi}{a} \right)^4 W_m \right\} \right.$$
$$\left. - P_{cr} \left(\frac{m\pi}{a} \right)^2 W_m \right] \sin \frac{m\pi x}{a} = 0$$

i.e.

$$W_{m,yyyy} - 2 \left(\frac{m\pi}{a} \right)^2 W_{m,yy} + \left(\frac{m\pi}{a} \right)^4 W_m - \left(\frac{m\pi}{a} \right)^2 \frac{P_{cr}}{D} W_m = 0 \quad \text{for any } m.$$

The solution of this ordinary differential equation is given by

$$W_m = A_m \cosh \alpha y + B_m \sinh \alpha y + C_m \cos \beta y + D_m \sin \beta y$$

where A_m to D_m are undetermined constants, and

$$\alpha = \sqrt{\left[\frac{m\pi}{a} \sqrt{\frac{P_{cr}}{D}} + \left(\frac{m\pi}{a} \right)^2 \right]} \qquad \beta = \sqrt{\left[\frac{m\pi}{a} \sqrt{\frac{P_{cr}}{D}} - \left(\frac{m\pi}{a} \right)^2 \right]}.$$

The conditions

$$W_m = W_{m,y} = 0 \text{ at } y = \pm b/2$$

lead to

$$A_m \cosh \frac{\alpha b}{2} + C_m \cos \frac{\beta b}{2} = 0$$

$$\alpha A_m \sinh \frac{\alpha b}{2} - \beta C_m \sin \frac{\beta b}{2} = 0$$

$$B_m \sinh \frac{\alpha b}{2} + D_m \sin \frac{\beta b}{2} = 0$$

$$\alpha B_m \cosh \frac{\alpha b}{2} + \beta D_m \cos \frac{\beta b}{2} = 0$$

where the decoupling of the y-symmetric and the y-antisymmetric buckled mode shapes can be seen. Subsequent steps are similar to those of the vibration problem in Example 6.2; thus, one obtains the values of P_{cr} corresponding to the roots of the characteristic equation. The lowest P_{cr} and the corresponding mode shape have to be identified only by trial and error (see Problem 3).

7.3 Static Flexure

The analysis for static deflections due to combined in-plane and transverse loading is illustrated below with reference to a simple problem.

Example 7.3 Combined Bending and Stretching of a Simply Supported Plate
    The simplest problem in this category is that of a uniaxially stressed simply supported plate subjected to transverse sinusoidal loading

$$q = q_{11} \sin \frac{\pi x}{a} \sin \frac{\pi y}{b}$$

Taking $a = b$ and the in-plane force to be N_x, the solution for Eq. (7.7) can be taken as

$$w = W_{11} \sin \frac{\pi x}{a} \sin \frac{\pi y}{a}$$

to yield

$$W_{11} = \frac{q_{11}}{\frac{\pi^2}{a^2}\left[\frac{4\pi^2 D}{a^2} + N_x\right]}$$

Expressed in terms of the counterpart without the in-plane force, the above equation becomes

$$w_{max} = \frac{w_{max \text{ without } N_x}}{\left[1 + \frac{N_x a^2}{4\pi^2 D}\right]}$$

Noting that the uniaxial compressive buckling load P_{cr} for the square plate is $\frac{4\pi^2 D}{a^2}$, the above result can be expressed as

$$w_{max} = \frac{w_{max \ without \ N_x}}{\left[1 + \frac{N_x}{P_{cr}}\right]}$$

This equation clearly shows that the influence of the in-plane force on the transverse deformation of the plate—stiffening it when tensile and slackening it when compressive—would be significant if its magnitude is a fairly large fraction of the corresponding critical value. Though this has been arrived at with reference to the above simple problem, this is in general true for other in-plane force fields and transverse loads.

7.4 Free Vibrations

The equation governing free vibrations of a plate subjected to in-plane forces is given by

$$D\nabla^4 w + \bar{\rho} w_{,tt} = N_x w_{,xx} + N_y w_{,yy} + 2N_{xy} w_{,xy} \qquad (7.8)$$

This equation admits of a Navier-type solution for the case of a simply supported plate without in-plane shear forces N_{xy}. Such a solution is illustrated below.

Example 7.4 Fundamental Frequency of An Initially Stressed Rectangular Plate
Consider a simply supported rectangular plate ($0 \leq x, y \leq a, b, a > b$) subjected to uniform uniaxial compression $N_x = -P$. Applying Navier's method exactly as done for the initially unstressed plate (Example 6.1), one gets the frequency equation for the present case as

$$D\left\{\left(\frac{m\pi}{a}\right)^4 + 2\left(\frac{m\pi}{a}\right)^2\left(\frac{n\pi}{b}\right)^2 + \left(\frac{n\pi}{b}\right)^4\right\} - P\left(\frac{m\pi}{a}\right)^2 - \bar{\rho}\,\omega_{mn}^2 = 0$$

and hence the natural frequencies as

$$\omega_{mn}^2 = \frac{D}{\bar{\rho}}\left[\left(\frac{m\pi}{a}\right)^2 + \left(\frac{n\pi}{b}\right)^2\right]^2 - \frac{P}{\bar{\rho}}\left(\frac{m\pi}{a}\right)^2 \quad \text{for } m, n = 1, 2, \text{ etc.}$$

with mode shapes having the corresponding number of half-waves in the two directions.

Comparing this with the counterpart for the initially unstressed plate, one finds, besides a decrease in all the frequencies as expected, a change in the fundamental mode itself. While the lowest ω_{mn} always corresponds to $m = n = 1$ for the unstressed case, here n alone is 1, while m depends on P as well as the aspect ratio of the plate. For example, if one considers a plate with $a = 5b$,

$$\omega_{m1}^2 = \frac{D\pi^4}{\bar{\rho}b^4}\left[\left(\frac{m}{5}\right)^2 + 1\right]^2 - \frac{P\pi^2}{\bar{\rho}b^2}\left(\frac{m}{5}\right)^2$$

Rewriting P in terms of the critical load P_{cr} equal to $\frac{4\pi^2 D}{b^2}$ as

$$P = \eta P_{cr} = \eta\frac{4\pi^2 D}{b^2}$$

one has

$$\omega_{m1}^2 = \frac{D\pi^4}{\bar{\rho}b^4}\left[\left\{\left(\frac{m}{5}\right)^2 + 1\right\}^2 - 4\eta\left(\frac{m}{5}\right)^2\right]$$

Considering the extreme case of $\eta \approx 1$, this can be expressed as

$$\omega_{m1}^2 \approx \frac{D\pi^4}{\bar{\rho}b^4}\left\{\left(\frac{m}{5}\right)^2 - 1\right\}^2$$

obviously becoming a minimum and nearly zero for $m = 5$, the shape corresponding to buckling of the plate into square panels. When $\eta \approx 0$, $m = 1$ yields the lowest frequency. In other words, as the compressive load P is increased from zero to the maximum possible value P_{cr}, the fundamental frequency decreases continuously from the value for the initially unstressed plate to zero, and with the mode shape changing from $m = 1$ to $m = 5$ through the intermediate values 2, 3 and 4 while n remains 1.

Such possible changes in the fundamental mode shape have to be kept in mind while solving rectangular plate problems involving initial stresses using approximate trial function approaches such as Rayleigh-Ritz method or Galerkin's method (see Chap. 8).

Summary

The governing equations for combined bending and stretching have been presented. Simple solutions have been illustrated for the analysis of the critical buckling loads and also for static flexure and free vibration in the presence of in-plane loads. Only uniform in-plane force fields have been considered in this chapter—the more complicated case of non-uniform force fields will be dealt with in Chap. 13.

Problems

1. With reference to the figure, sketch the nodal lines after the plate buckles. What is the percentage change in the buckling load due to

   (a) 20% increase in E?
   (b) 10% decrease in both a & b?
   (c) 10% increase in thickness?
   (d) 100% increase in a with b unchanged?

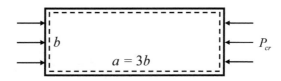

Problem 7.1

2. Corresponding to Case(a) of Example 7.1, determine the buckling load for $a/b = \sqrt{2}, \sqrt{6}$ and $\sqrt{12}$. From these results, deduce that, for $a/b > \sqrt{12}$, the buckling load can be taken as $(4\pi^2 D/b^2)$ with less than 2% error.

3. For the square plate of Example 7.2, show that $P_{cr} = 8.604\frac{\pi^2 D}{a^2}$ corresponds to a y-symmetric mode with $m = 1$, and $P_{cr} = 7.691\frac{\pi^2 D}{a^2}$ to a y-symmetric mode with $m = 2$. Sketch the y-variation of these shapes. (Note that $m = 2$ yields a lower value of P_{cr}—this is actually the lowest buckling load for the SCSC square plate).

4. With reference to Case(a) of Example 7.1, assuming that the plate is supported by an elastic foundation of $k_F = \frac{4\pi^4 D}{b^4}$, show that the first three transition points as in Fig. 7.3 occur at $a/b = 0.946, 1.64$ and 2.32.

Chapter 8
Approximate Solutions

The solutions presented so far, for statics, dynamics and stability, are all exact in the sense that they rigorously satisfy the governing equations and the boundary conditions. However, such exact solutions are not possible for plates of arbitrary geometry and boundary conditions. Hence, one has to think in terms of alternative solution methodologies which, though approximate, would yield results acceptable for purposes of engineering analysis. This chapter presents a brief discussion of a few such approximate solutions.

8.1 Analytical and Numerical Methods

Approximate methods can broadly be classified into two categories—*analytical methods* and *numerical methods*. In analytical methods, the unknown quantities (deflections, stresses, etc.) are sought in the form of functions of the coordinate variables, while in numerical methods, one seeks the numerical values of these quantities at many chosen locations of the domain. Both these categories of methods have their own advantages—while analytical solutions enable one to immediately realize how the final result depends on the various parameters involved, this is not possible from numerical solutions unless one solves the problem repeatedly with different input values for each parameter; however, analytical methods are not applicable for all problems, and in this respect, numerical methods hold sway as they enable the engineer to solve many a complicated problem which would otherwise remain unsolved.

Examples of analytical methods include the techniques of Rayleigh–Ritz, Galerkin, Kantorovich, collocation and least-squares, among others. Popular among the numerical methods are the finite difference method, finite element method and the boundary element method.

© The Author(s) 2021
K. Bhaskar and T. K. Varadan, *Plates*,
https://doi.org/10.1007/978-3-030-69424-1_8

A complete discussion of all the approximate methods is beyond the scope of the present book. Only two methods—those of Rayleigh–Ritz and Galerkin—are explained here as applied to the study of statics, dynamics as well as stability of rectangular plates.

8.2 Rayleigh–Ritz Method

8.2.1 Static Flexure

For problems of static loading, Rayleigh–Ritz method is based on the principle of minimum potential energy, stated as follows:

> Of all displacement configurations which satisfy compatibility and kinematic boundary conditions, the one corresponding to stable equilibrium has the least potential energy.

Compatible displacement configurations are those that preserve the continuum of the structure, i.e. those due to which the deformed structure would be free of any voids or discontinuities. *Kinematic* (or *geometric* or *essential*) boundary conditions are those specified in terms of the deflection or the slope of the plate; they are to be distinguished from the *natural* (or *force*) boundary conditions which are specified in terms of the shear force or the bending moment. Displacement functions that satisfy compatibility as well as kinematic boundary conditions are called *admissible* functions. Corresponding to any such admissible function, the deformed structure has a certain potential energy comprising of the strain energy of deformation and the potential energy of the applied loads. Thus, the determination of the actual deformed state of the structure via the principle of minimum potential energy amounts to the identification, from the *infinite set* of admissible configurations, of that corresponding to minimum potential energy.

If, instead of considering all the infinite possible configurations, one confines attention to a convenient subset, and identifies, from it, the one corresponding to minimum potential energy, then the result is an approximation to the actual deformed state. This is the essence of Rayleigh–Ritz method.

For a plate, the strain energy of deformation is given by (see Sect. 2.8)

$$U_e = \frac{D}{2} \int \left[\left(\nabla^2 w \right)^2 - 2(1-\mu)(w_{,xx}w_{,yy} - w_{,xy}^2) \right] \mathrm{dArea} \qquad (8.1)$$

As the plate deforms, the loads acting on it (comprising of the following transverse loads as a most general case: a distributed load of intensity $q(x,y)$, a set of line loads denoted by Q_i acting along the lines denoted by Γ_i, and a set of concentrated point loads P_i acting at the points (x_i, y_i), all of them taken as positive when acting in the $+z$ direction) move through a distance; with reference to the undeformed configuration as the datum, their final potential energy is defined as

$$V = -\iint qw \, dx \, dy - \int_{\Gamma_i} Q_i w|_{\Gamma_i} d\Gamma - \sum_i P_i w|_{(x_i, y_i)} \qquad (8.2)$$

While using the above equation, the support reactions should also be included under loads; however, the contribution of the support reactions to the potential energy becomes zero for all the three classical boundary conditions (simple or fixed or free). Further, if the loads include some moments, the work done by them as they rotate through an angle should be considered. It should be noted that the net potential energy V of all the loads, as given by Eq. (8.2), is always negative while some of the terms can be positive.

The total potential energy Π is given by

$$\Pi = U_e + V \qquad (8.3)$$

The application of Rayleigh–Ritz method to problems of static flexure is illustrated below. In all these problems, μ is taken to be 0.3.

Example 8.1 Simply supported square plate under sinusoidal loading.

As a first case, a problem for which the exact solution is possible is taken up so that the accuracy of the Rayleigh–Ritz method can be studied. A square plate $(0 \le x, y \le a)$ simply supported at all edges and subjected to transverse load given by

$$q = q_o \sin \frac{\pi x}{a} \sin \frac{\pi y}{a}$$

is considered. The boundary conditions are split into kinematic boundary conditions and natural boundary conditions as:

Kinematic conditions: $w = 0$ at $x = 0, a$ and $y = 0, a$

Natural conditions: $\begin{array}{l} M_x = 0 \quad \text{at } x = 0, \ a \\ M_y = 0 \quad \text{at } y = 0, \ a. \end{array}$

A solution, satisfying the kinematic boundary conditions alone, is chosen as

$$w = Ax(x - a)y(y - a)$$

where A is an undetermined constant. This equation represents a family of admissible functions with different values of A; Rayleigh–Ritz method can now be used to identify, from among these, the admissible function (i.e. the specific value of A) corresponding to the lowest potential energy.

The energies, for the assumed w, turn out to be

$$U_e = \frac{11}{45} Da^6 A^2 \quad V = -\frac{16}{\pi^6} q_o a^6 A$$

and hence,

$$\Pi = \frac{11}{45}Da^6A^2 - \frac{16}{\pi^6}q_oa^6A$$

Corresponding to minimum Π, $\frac{d\Pi}{dA} = 0$ and $\frac{d^2\Pi}{dA^2} > 0$.

By equating $\frac{d\Pi}{dA}$ to zero, one gets $A = \frac{360q_o}{11\pi^6D}$.

Differentiating further, it is easily seen that $\frac{d^2\Pi}{dA^2}$ is positive. This will be true for all Rayleigh–Ritz solutions for problems of static loading; one need not verify this.

Thus, the approximate solution, corresponding to the chosen family of functions, is

$$w = \frac{360q_o}{11\pi^6D}x(x-a)y(y-a)$$

while the exact solution, which can be obtained by Navier's method (Sect. 4.2), is

$$w = \frac{q_oa^4}{4\pi^4D}\sin\frac{\pi x}{a}\sin\frac{\pi y}{a}$$

Comparing the maximum deflections, at $x = y = a/2$, one gets

$$w_{max}(\text{approx.}) = 0.002128\frac{q_oa^4}{D}$$
$$w_{max}(\text{exact}) = 0.002566\frac{q_oa^4}{D}$$

i.e. an error given by

$$\frac{w_{approx} - w_{exact}}{w_{exact}} = -0.1707 \quad \text{or} \quad -17.07\%$$

If one compares the maximum M_x, again at the centre, one gets

$$M_{x\ max} = -D\left(w_{,xx} + \mu w_{,yy}\right)\big|_{x=y=a/2}$$
$$= \begin{cases} 0.02213q_oa^2 \ (\text{approx.}) \\ 0.03293q_oa^2 \ (\text{exact}) \end{cases}$$

i.e. an error of -32.8%.

To improve the accuracy of the results, one can think in terms of a multi-term approximation of w. For example, a three-term solution can be assumed as

$$w = Ax(x-a)y(y-a) + Bx\left(x - \frac{a}{4}\right)\left(x - \frac{3a}{4}\right)(x-a)y(y-a)$$
$$+ Cx(x-a)y\left(y - \frac{a}{4}\right)\left(y - \frac{3a}{4}\right)(y-a)$$

Fig. 8.1 Normalized plots

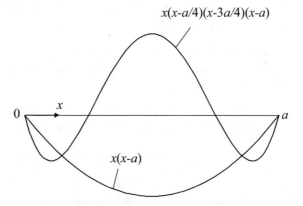

$x(x-a/4)(x-3a/4)(x-a)$

$x(x-a)$

which represents a larger family of admissible functions than earlier, now depending on the values of A, B and C. The shapes associated with the above approximation can be visualized from the plots of Fig. 8.1, where one should note that the symmetry of the actual deformation has been accounted for.

The potential energy Π will now depend on A, B and C. For minimum Π, one has

$$\frac{\partial \Pi}{\partial A} = \frac{\partial \Pi}{\partial B} = \frac{\partial \Pi}{\partial C} = 0$$

which reduce to three linear algebraic equations in A, B and C.

Solving them, one gets

$$A = 0.03662\frac{q_o}{D}, \quad B = C = -0.03375\frac{q_o}{a^2 D}$$

corresponding to which

$$w_{max} = 0.002552\frac{q_o a^4}{D} \quad (-0.55\% \text{ error})$$
$$M_{x\,max} = 0.03203 q_o a^2 \quad (-2.73\% \text{ error })$$

If the chosen family of functions includes the exact state of deformation as well, then Rayleigh–Ritz method would automatically yield the exact solution. This can easily be verified by starting with

$$w = A \sin \frac{\pi x}{a} \sin \frac{\pi y}{a}$$

from which A will be obtained as

$$A = \frac{q_o a^4}{4\pi^4 D}$$

This coincides with the exact solution.

The following example deals with the problem of a clamped plate for which the exact solution is cumbersome.

Example 8.2 Clamped square plate under uniformly distributed load.

The exact solution for this problem can be obtained by superposition of the solutions of

(a) a simply supported plate under uniform load;
(b) a simply supported plate loaded by edge moments at the edges $y = $ constant; and
(c) a simply supported plate loaded by edge moments at the edges $x = $ constant.

Levy-type solutions for (a) and (b) are obtained by using a sine series in the x-direction while that for (c) is in terms of a sine series in the y-direction. After superposition, when the zero slope conditions are enforced at the edges, the edge moments can be determined. This involves the solution of an infinite number of algebraic equations in terms of the Fourier coefficients of the moments; by appropriate truncation, one can obtain results of any desired accuracy.

Such problems, though amenable to exact analysis, are preferably solved by using an approximate method which yields a reasonably accurate result without tedious effort. The Rayleigh–Ritz solution for the problem of the clamped plate is given below.

For a plate in the domain $-a \leq x, y \leq a$, the following admissible functions are compared here.

$$\text{I: } w = A(x^2 - a^2)^2(y^2 - a^2)^2$$

$$\text{II : } w = A\left(1 + \cos\frac{\pi x}{a}\right)\left(1 + \cos\frac{\pi y}{a}\right)$$

Both these functions satisfy the zero deflection and slope conditions at all the edges and also account for the symmetry of deformation about the lines $x = 0$ and $y = 0$.

The results for $w_{\max}$ corresponding to these functions are:

$$\text{(I) } 0.02127\frac{qa^4}{D}, \quad \text{(II) } 0.02053\frac{qa^4}{D}$$

while the exact solution is $0.02016\frac{qa^4}{D}$.

These results show that the second function is superior, and that a very good estimate of the maximum deflection can be obtained by using a simple one-term approximation for w.

An important advantage of Rayleigh–Ritz method is the ease with which one can account for changes in thickness or material properties over the domain; this is illustrated in the following example.

Fig. 8.2 Plate of
non-uniform thickness

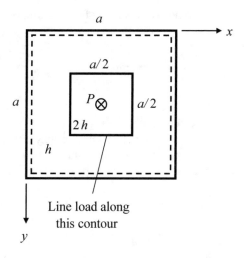

Example 8.3 A plate of non-uniform thickness.

Consider a simply supported square plate as shown in Fig. 8.2, where the central square portion is two times thicker than the outer portion. The plate is subjected to a central concentrated load P as well as another load, of total magnitude P again, but distributed uniformly along the boundary of the inner square.

Noting the symmetry of the problem about the lines $x = a/2$ and $y = a/2$, a one-term solution can be sought as

$$w = A \sin \frac{\pi x}{a} \sin \frac{\pi y}{a}$$

It is sufficient to find out U_e and V for one quarter of the domain (say, $0 \leq x, y \leq a/2$) because of the double symmetry. The flexural rigidity for the inner portion is eight times that for the outer portion; if $D_{\text{outer}} = D$, then $D_{\text{inner}} = 8D$.

Thus, for one quarter of the plate,

$$U_e = \frac{D}{2} \int\limits_{y=0}^{a/2} \int\limits_{x=0}^{a/4} \left[(\nabla^2 w)^2 - 2(1-\mu)(w_{,xx} w_{,yy} - w_{,xy}^2) \right] dx\, dy$$

$$+ \frac{D}{2} \int\limits_{y=0}^{a/4} \int\limits_{x=a/4}^{a/2} \left[(\nabla^2 w)^2 - 2(1-\mu)(w_{,xx} w_{,yy} - w_{,xy}^2) \right] dx\, dy$$

$$+ \frac{8D}{2} \int\limits_{y=a/4}^{a/2} \int\limits_{x=a/4}^{a/2} \left[(\nabla^2 w)^2 - 2(1-\mu)(w_{,xx} w_{,yy} - w_{,xy}^2) \right] dx\, dy$$

For calculation of the potential energy V of the applied loads, one should note that the central concentrated load corresponding to one quarter of the domain is $P/4$. Further, the intensity of the distributed line load is $P/2a$. Corresponding to these loads, the potential energy of one quarter of the plate can be obtained as

$$V = -\frac{P}{4}w\Big|_{x=y=a/2} - \int_{a/4}^{a/2}\frac{P}{2a}w\Big|_{y=a/4}\,dx - \int_{a/4}^{a/2}\frac{P}{2a}w\Big|_{x=a/4}\,dy$$

Proceeding further, one gets $A = 0.004070\frac{Pa^2}{D}$.

Note that a one-term solution may not be adequate for problems of this kind; for instance, if one needs accurate estimates of stresses at the boundary of the inner square here, a multi-term approximation is certainly called for.

8.2.2 Buckling

A state of equilibrium for a stressed body can in general be stable, unstable or neutral. When subjected to a small disturbance, if the body has a tendency to regain its original position, it is said to be in stable equilibrium; if it has a tendency to diverge further away from the original position, it is under unstable equilibrium; and, if neither of these tendencies is present and the body simply stays put in the disturbed position, it is said to be in neutral equilibrium.

For a plate subjected to in-plane loads below their critical values (i.e. too small to cause buckling), it is clear that the state of equilibrium is stable, because any transverse displacement arising due to a small disturbance dies out. If the loads are greater than their critical values, then any small disturbance would lead to larger and larger deflections eventually leading to failure of the plate; thus the state of equilibrium is unstable. When the loads are exactly equal to the critical values, the state of equilibrium is neutral because the disturbed plate has no tendency either to return to the original flat position or to diverge away from it.

In terms of the potential energy, a state of stable equilibrium corresponds to minimum energy while that of unstable equilibrium to maximum energy; for neutral equilibrium, the energy is the same for all adjacent configurations. Hence, the energy criterion for buckling can be stated as:

> Among the various values of in-plane loads, those corresponding to buckling are the ones for which the potential energy of the perturbed plate configuration is stationary.

This implies that the first and the second variations of the potential energy are zero, or, in other words, that the potential energy of the buckled plate would be zero with reference to the compressed flat configuration that is about to buckle.

Consider a plate, initially undeformed, subjected to certain in-plane loads resulting in an in-plane force field specified in terms of $N_x(x,y)$, $N_y(x, y)$ and $N_{xy}(x, y)$. Let the critical values of these be $N_{xcr}(x, y)$, $N_{y\,cr}(x, y)$ and $N_{xy\,cr}(x, y)$ corresponding to the onset of buckling. At this stage, just before buckling, the flat plate has some membrane strain energy due to in-plane deformation, and the in-plane loads have lost some potential energy due to the work done by them. These, however, are not of interest in the present context; what is required is to determine the potential energy of the buckled plate with the prebuckled state taken as the datum. This can be correctly derived by using the following assumptions:

(a) During the transition of the plate from the compressed prebuckled state to the adjacent bent state, the in-plane force field described by $N_{x\,cr}(x, y)$, $N_{y\,cr}(x, y)$ and $N_{xy\,cr}(x, y)$ remains unchanged.

(b) Further, during this small deformation, the x and y displacements of any point of the mid-plane are negligible.

One readily apparent component of the potential energy of the bent state with reference to the prebuckled flat state is the strain energy due to bending and twisting, already derived earlier and given by Eq. (8.1) in terms of w. Further, associated with this bending and twisting which are taken to occur within the purview of assumption (b) above, the mid-plane undergoes stretching and shearing; thus, the membrane strains undergo small changes with reference to their prebuckling values while the membrane forces remain unchanged, and correspondingly there is a change in the membrane strain energy. The third component to be considered, namely the change in potential energy of the applied in-plane loads, is zero because these loads do not move in their own directions by virtue of assumption (b).

The change in the membrane strain energy associated with bending and twisting can be calculated as the sum of the work done by the unvarying in-plane forces during this deformation, or more specifically, by $N_{xcr}(x, y)$ during the x-direction stretch, by $N_{ycr}(x, y)$ during the y-direction stretch, and by $N_{xycr}(x, y)$ during the mid-plane shear distortion. This is as explained below.

Consider a small element ABCD of the mid-plane of the critically stressed plate as in Fig. 8.3, with N_{xcr}, N_{ycr} and N_{xycr} shown in their positive directions. To find the stretch in the x-direction, consider the line segment AB. As the plate buckles, AB moves to the position A'B' as shown in Fig. 8.4. Thus, the x-direction stretch is given by

$$\sqrt{dx^2 + (w_{,x}dx)^2} - dx \approx \frac{1}{2}w_{,x}^2 dx$$

The corresponding work done by the unvarying $N_{xcr}(x, y)$ is $\frac{1}{2}N_{x\,cr}w_{,x}^2 dx\,dy$ for the element ABCD.

Similarly, corresponding to y-direction stretch, one gets $\frac{1}{2}N_{y\,cr}w_{,y}^2 dx\,dy$.

The in-plane shear strain due to buckling is obtained as shown in Fig. 8.5; let A'B'C'D' be the deformed plate element, shown such that the slopes in both x and y directions are positive at A'. B'' and D'' be the projections of B' and D', respectively on the plane parallel to the x–y plane and passing through A'. Thus, $\angle B'A'B'' = w_{,x}$ and $\angle D'A'D'' = w_{,y}$. To find the shear strain of the element, let us consider the deviation from $90°$ of one of its subtended angles, say $\angle D'A'B'$, noting that the shear strain would be positive when this angle is less than $90°$.

Fig. 8.3 Critically stressed
plate before buckling

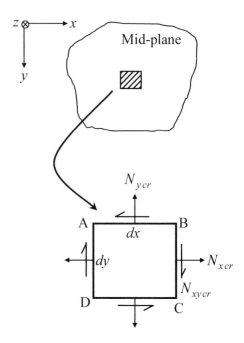

Fig. 8.4 Elongation in the
x-direction during buckling

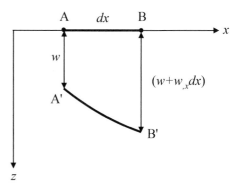

One can visualize the final deformed position A′B′C′D′ as resulting from the following four actions:

(a)   rigid body translation of the element ABCD in the z-direction such that A moves to A′, B to B″, D to D″ and similarly C to a point (not shown in Fig. 8.5) vertically below it in the plane of A′, B″ and D″;

(b)   rigid body rotation of the element about A′D″ so that B″ moves to B′ and with ∠D″A′B′ = 90°;

(c)   rigid body rotation of the element about A′B′ so that D″ moves to E in the plane of A′, B′ and D′, and with ∠B′A′E = 90°;

(d)   shear distortion due to which E moves to D′.

Fig. 8.5 Determination of shear strain

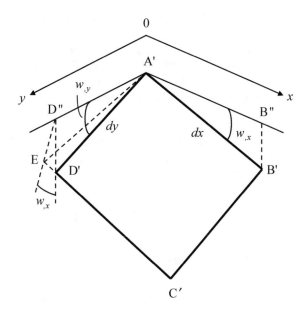

Thus, the shear strain is given by $\angle EA'D'$. Considering the two right-angled triangles $\triangle A'D''D'$ and $\triangle D'ED''$, one gets

$$D'E = (D'D'')w_{,x} = (A'D')w_{,y}w_{,x} = w_{,x}w_{,y}dy$$

and hence, $\angle EA'D' = \frac{D'E}{dy} = w_x w_y$. Thus, the work done by the shear forces during buckling is $N_{xy\,cr}w_{,x}w_{,y}dx\,dy$ for the element ABCD.

Adding the above three components and integrating over the area of the mid-plane, one gets the total change in membrane strain energy associated with the transition from the flat prebuckled state to the adjacent bent configuration.

Thus, the total potential energy of the buckled plate with respect to the prebuckled flat position as the datum can finally be expressed as

$$\Pi = \frac{D}{2}\int[(\nabla^2 w)^2 - 2(1-\mu)(w_{,xx}w_{,yy} - w_{,xy}^2)]dx\,dy$$
$$+ \frac{1}{2}\iint[N_{x\,cr}w_{,x}^2 + N_{y\,cr}w_{,y}^2 + 2N_{xy\,cr}w_{,x}w_{,y}]\,dx\,dy \tag{8.4}$$

It should be noted that the above potential energy is in terms of the internal forces $N_{xcr}(x, y)$, $N_{y\,cr}(x, y)$ and $N_{xy\,cr}(x, y)$ and not the edge loads themselves. (It should be noted that the general governing equation for combined bending and stretching [Eq. (7.7)] is also in terms of the internal force field.) When the loads happen to be uniform along the edges, the internal forces also turn out to be uniform with the corresponding values, and hence can be used directly with an assumed buckled shape to obtain their critical values by Rayleigh–Ritz method. In the case of non-uniform edge loads, however, the internal force field should first be found out by appropriate plane stress analysis and this forms a preliminary step for the subsequent buckling analysis. (Such problems will be discussed in Chap. 13.)

The following examples serve to illustrate the application of Rayleigh–Ritz method for the study of buckling due to uniform in-plane loads.

Example 8.4 SCSC square plate under uniaxial compression (Fig. 8.6).

From earlier experience, it is known that the lowest buckling load corresponds to a y-symmetric mode with no nodal lines parallel to the x-axis. Thus, an appropriate choice for the mode shape is

$$w = A \sin \frac{m \pi x}{a} \left(1 - \cos \frac{2\pi y}{a}\right)$$

which satisfies all the edge conditions.

Substitution for w in the expression for the potential energy leads to

$$\Pi = \frac{(3m^4 + 8m^2 + 16)D\pi^4 A^2}{8a^2} + \frac{3\pi^2 m^2 N_{x\,\mathrm{cr}} A^2}{8}$$

For stationary potential energy,

$$\frac{d\Pi}{dA} = \left[\frac{(3m^4 + 8m^2 + 16)D\pi^4}{4a^2} + \frac{3\pi^2 m^2 N_{x\,\mathrm{cr}}}{4}\right]A = 0$$

which, for a non-trivial solution, implies that

Fig. 8.6 Buckling of SCSC square plate

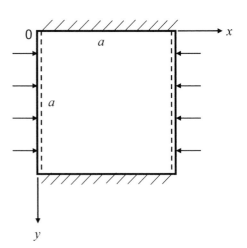

$$\frac{(3m^4 + 8m^2 + 16)D\pi^4}{4a^2} + \frac{3\pi^2 m^2 N_{x\,cr}}{4} = 0$$

i.e. $N_{x\,cr} = \frac{-(3m^4 + 8m^2 + 16)D\pi^2}{3a^2 m^2}$

It is easy to see that the above condition would also make $\frac{d^2\Pi}{dA^2}$ zero.

Thus, for a buckling problem, once the first variation of the potential energy is enforced to be zero, the solution would automatically satisfy the condition of neutral equilibrium.

It is clear that N_{xcr} has an infinite number of values, corresponding to $m = 1, 2$, etc., and that the lowest buckling load has to be identified by trial and error. We have

$$N_{x\,cr} = -\frac{9\pi^2 D}{a^2} \quad \text{for } m = 1$$

$$= -\frac{8\pi^2 D}{a^2} \quad \text{for } m = 2$$

$$= -\frac{12.26\pi^2 D}{a^2} \quad \text{for } m = 3, \text{ and so on.}$$

Thus, the lowest buckling load has a magnitude of $\frac{8\pi^2 D}{a^2}$ with the following mode shape.

$$w = A \sin \frac{2\pi x}{a} \left(1 - \cos \frac{2\pi y}{a}\right)$$

Compared to the exact solution ($\frac{7.691\pi^2 D}{a^2}$ from Example 7.2), the lowest buckling load is thus over-predicted by 4.02%. The error can be reduced by using a multi-term approximation, which would simultaneously enable one to obtain the subsequent buckling loads also. This will be clear from Example 8.5.

In the foregoing example, it was seen that the buckling load is over-predicted by the energy method. This is an important feature of this method and is true for all structures for which it is used. This is so because the approximation of the mode shape implies that the structure is constrained to deform accordingly, and is thus made stiffer. As the buckling load of a structure depends directly on its stiffness, the stiffer approximation of the Rayleigh–Ritz method leads to a higher estimate of the actual buckling load. (This is true for natural frequencies also as will be seen later.)

Example 8.5 Shear buckling of a rectangular plate.

Consider a simply supported plate ($0 \le x, y \le a, b$) subjected to shear forces N_{xy} alone. This is a case for which an exact solution is not possible. Attempting a one-term Rayleigh–Ritz solution as

$$w = A \sin \frac{m\pi x}{a} \sin \frac{n\pi y}{b}$$

and proceeding ahead, one would find that the methodology does not lead to a buckling load, but simply to the trivial solution given by $A = 0$ for all m and n. This means that none of the buckled shapes can be described in variables-separable form as a product of a function of x alone and a function of y alone. Hence one has to start with a more general approximation.

For example, using the four-term approximation

$$
\begin{aligned}
w = {} & A_{11} \sin \frac{\pi x}{a} \sin \frac{\pi y}{b} + A_{12} \sin \frac{\pi x}{a} \sin \frac{2\pi y}{b} \\
& + A_{21} \sin \frac{2\pi x}{a} \sin \frac{\pi y}{b} + A_{22} \sin \frac{2\pi x}{a} \sin \frac{2\pi y}{b}
\end{aligned}
$$

Rayleigh–Ritz methodology leads to the following algebraic equations, wherein $\alpha = \frac{a}{b}$.

$$
\frac{\pi^4 D(1 + \alpha^2)^2}{4a^2\alpha} A_{11} - \frac{32 N_{xy}}{9} A_{22} = 0
$$

$$
\frac{\pi^4 D(1 + 4\alpha^2)^2}{4a^2\alpha} A_{12} + \frac{32 N_{xy}}{9} A_{21} = 0
$$

$$
\frac{\pi^4 D(4 + \alpha^2)^2}{4a^2\alpha} A_{21} + \frac{32 N_{xy}}{9} A_{12} = 0
$$

$$
\frac{4\pi^4 D(1 + \alpha^2)^2}{a^2\alpha} A_{22} - \frac{32 N_{xy}}{9} A_{11} = 0
$$

Here, the uncoupling of the constants A_{11} and A_{22} from A_{12} and A_{21} should be noted.

Seeking non-trivial solutions, one obtains

$$
N_{xy} = \pm \frac{9\pi^4 D(1 + \alpha^2)^2}{32a^2\alpha} \quad \text{with} \quad \frac{A_{22}}{A_{11}} = \pm\frac{1}{4} \quad \text{and} \quad A_{12} = A_{21} = 0
$$

and

$$
\pm \frac{9\pi^4 D(1 + 4\alpha^2)(4 + \alpha^2)}{128a^2\alpha} \quad \text{with} \quad \frac{A_{21}}{A_{12}} = \mp \frac{(1 + 4\alpha^2)}{(4 + \alpha^2)}
$$
$$
\text{and } A_{11} = A_{22} = 0
$$

Qualitatively, the difference between positive and negative N_{xy} is only one of direction, and hence it is sufficient to consider just the positive quantities. Thus, the above solution has led to two critical values of N_{xy}, and the lower of them corresponds to the first buckled shape of the plate.

One can easily verify that, irrespective of the value of α, the lowest buckling load is given by

$$
N_{xy\mathrm{cr}} = \frac{9\pi^4 D\left(1 + \alpha^2\right)^2}{32a^2\alpha}
$$
i.e. $\frac{11.10\pi^2 D}{a^2}$ for a square plate.

The above result can be improved by taking more terms in the double Fourier series for w; such a procedure would yield a converged value of $\frac{9.34\pi^2 D}{a^2}$ for a square plate.

8.2.3 Free Vibration Analysis

Rayleigh–Ritz approach for the analysis of free vibrations of a structure is based on Lagrange's equations of motion given by

$$\frac{\mathrm{d}}{\mathrm{d}t}\left(\frac{\partial T}{\partial \dot{q}_i}\right) + \frac{\partial U_e}{\partial q_i} = 0 \quad \text{with } i = 1, 2, \ldots, N \tag{8.5}$$

where q_i are the N degrees of freedom of the structure, t is time, T and U_e are kinetic and strain energies, respectively, and an overdot denotes a time-derivative.

The essence of Rayleigh–Ritz approach is to approximate the continuous system having infinite degrees of freedom as a finite-degree-of-freedom system. For a plate, this is achieved by approximating the deflection function $w(x, y, t)$ through the use of an expansion in terms of known admissible space functions with unknown time-dependent amplitudes. These unknown amplitudes comprise the finite degrees of freedom and correspond to q_i in Lagrange's equations.

Before proceeding ahead, it is necessary to obtain the expression for the kinetic energy T in terms of the deflection function.

For a general three-dimensional body, the kinetic energy can be expressed in terms of the displacement fields u, v and w, along the directions x, y and z, respectively as

$$T = \frac{1}{2} \int \rho\left(\dot{u}^2 + \dot{v}^2 + \dot{w}^2\right)\mathrm{d}\mathrm{Vol} \tag{8.6}$$

where ρ is the mass density.

Using the assumed displacement field of CPT [Eq. (2.4)], the above equation can be rewritten as

$$T = \frac{1}{2} \int\limits_{z=-h/2}^{h/2} \int \rho(z^2 \dot{w}_{,x}^2 + z^2 \dot{w}_{,y}^2 + \dot{w}^2)\mathrm{d}\mathrm{Area}\,\mathrm{d}z \tag{8.7}$$

Noting that $\int \rho z^2 dz$ denotes the mass moment of inertia of an element of unit area of the plate about the mid-plane, and that $\dot{w}_{,x}$ and $\dot{w}_{,y}$ are the rates of rotation of the element, one can see that the first two terms correspond to rotational energy of the plate. For thin plates falling under the purview of CPT, this rotational energy is negligibly small and hence can be ignored as compared to the third term representing the energy of translation in the transverse direction. (This is consistent with the neglect of rotary inertia while deriving the equation of motion in Chap. 6.)

Thus, the kinetic energy becomes

$$T = \frac{1}{2} \int \bar{\rho} \dot{w}^2 \mathrm{d}\text{Area} \tag{8.8}$$

where $\bar{\rho}$ is the mass per unit area of the plate.

The application of Rayleigh–Ritz to some free vibration problems is illustrated in the following examples.

Example 8.6 Determination of the fundamental frequency of a SCSC square plate.

A square plate ($0 \leq x, y \leq a$), simply supported at $x = 0$ and $x = a$, and clamped at the other two edges is considered. It is known that the fundamental mode of vibration for this plate would be such that there are no nodal lines in the x or y direction. Keeping this in mind, one can seek a solution for w as

$$w = A(t) \sin \frac{\pi x}{a} \left(1 - \cos \frac{2\pi y}{a} \right)$$

which satisfies all the edge conditions.

The energies turn out to be

$$U_e = \frac{27\pi^4 D A^2}{8a^2} \quad T = \frac{3\bar{\rho}a^2 \dot{A}^2}{8}$$

Substitution into the Lagrange's equation given by

$$\frac{d}{dt}\left(\frac{\partial T}{\partial \dot{A}}\right) + \frac{\partial U_e}{\partial A} = 0$$

yields, after simplification,

$$\ddot{A} + \frac{9\pi^4 D}{\bar{\rho}a^4} A = 0$$

Hence the natural frequency is given by

$$\omega^2 = \frac{9\pi^4 D}{\bar{\rho}a^4}$$

$$\text{i.e.} \quad \frac{\overline{\rho}\omega^2 a^4}{D} = 9\pi^4 = 876.68$$

as against the exact value of 838.15 (Example 6.2).

Thus, the natural frequency is overestimated by Rayleigh–Ritz method. This is due to the stiffening effect of the assumed mode shape as has been explained earlier with regard to the estimation of the buckling load.

If one is interested in improving the accuracy of the approximate solution, more terms have to be taken in the assumed mode shape. In the present case, the variation of w in the x-direction has been assumed as a half-sine wave and this coincides with the exact mode shape. Thus, only the y-variation has to be approximated more realistically, and this can be done by adding a second term (again symmetric about $y = a/2$) as given below:

$$w = \sin \frac{\pi x}{a} \left[A(t)\left(1 - \cos \frac{2\pi y}{a}\right) + B(t)\left(\cos \frac{2\pi y}{a} - \cos \frac{4\pi y}{a}\right) \right]$$

which satisfies the clamped edge conditions at $y = 0,a$.

This leads to

$$U_e = \frac{D}{2}\left(657.51 A^2 - 1217.61 AB + 7646.61 B^2\right)$$
$$T = \frac{\overline{\rho}a^2}{2}\left(0.75\dot{A}^2 - 0.5\,\dot{A}\dot{B} + 0.5\dot{B}^2\right)$$

and hence,

$$0.75\overline{\rho}a^2\ddot{A} - 0.25\overline{\rho}a^2\ddot{B} + 657.51 DA - 608.81 DB = 0$$
$$-0.25\overline{\rho}a^2\ddot{A} + 0.5\overline{\rho}a^2\ddot{B} - 608.81 DA + 7646.61 DB = 0$$

$$\text{i.e. } \overline{\rho}a^2 \begin{bmatrix} 0.75 & -0.25 \\ -0.25 & 0.50 \end{bmatrix} \begin{Bmatrix} \ddot{A} \\ \ddot{B} \end{Bmatrix} + D \begin{bmatrix} 657.51 & -608.81 \\ -608.81 & 7646.61 \end{bmatrix} \begin{Bmatrix} A \\ B \end{Bmatrix} = \begin{Bmatrix} 0 \\ 0 \end{Bmatrix}$$

For harmonic vibrations, by assuming

$$\begin{Bmatrix} A \\ B \end{Bmatrix} = \begin{Bmatrix} \overline{A} \\ \overline{B} \end{Bmatrix} e^{i\omega t}$$

and proceeding further, one gets, for a non-trivial solution (i.e. $\overline{A}$ and $\overline{B}$ are not both zero),

$$\frac{\overline{\rho}\omega^2 a^4}{D} = 847.61 \quad \text{and} \quad 17{,}584.6$$

The first value corresponds to the fundamental frequency and the second to that of the mode symmetric about $y = a/2$ and with two nodal lines parallel to the x-axis. The exact values are 838.15 and 16,665.7 (see Example 6.2).

Thus the approximate solution now yields a more accurate estimate of the fundamental frequency compared to the one-term approximation. Further, a first estimate of another frequency is also obtained. Substitution of ω into any one of the two equations of motion yields the corresponding mode shape in terms of $(\overline{A}/\overline{B})$.

The following example illustrates the superiority of an analytical solution compared to a numerical solution for problems where the influence of certain parameters has to be studied.

Example 8.7 Maximization of the fundamental frequency of a clamped square plate.

The problem is posed thus: "A square plate $(-a \leq x, y \leq a)$, clamped all around, is of uniform thickness h_o. It is required to explore the possibility of increasing the fundamental frequency of the plate, without changing the weight, by employing a linear taper of the thickness in the x-direction alone. Symmetry of the plate about the y-axis has to be maintained. A Rayleigh–Ritz solution with the mode shape

$$w = A\left(1 + \cos\frac{\pi x}{a}\right)\left(1 + \cos\frac{\pi y}{a}\right)$$

may be considered adequate for this optimization study.

What the problem means is this: After calculating the fundamental frequency of the uniform plate using the suggested one-term Rayleigh–Ritz method, one should compare it with that of a plate for which the thickness varies linearly in the x-direction as shown in Fig. 8.7, while it remains uniform in the y-direction. (Note that h_2 need not necessarily be greater than h_1.) Using the condition that the weight of the tapered plate is equal to that of the original uniform plate, and finding out the fundamental frequency of the tapered plate as a function of the taper parameter h_2/h_1, one has to arrive at the best taper for obtaining the maximum frequency.

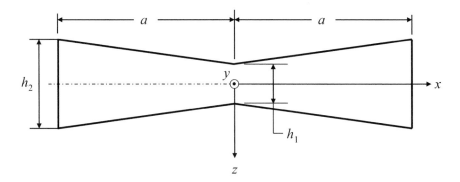

Fig. 8.7 Tapered plate

For the plates to have the same weight, the area of Fig. 8.7 should be equal to that of its counterpart for the uniform plate,

$$\text{i.e. } 2\frac{(h_1 + h_2)}{2}a = 2ah_o \quad \text{or} \quad h_1 + h_2 = 2h_o$$

Denoting h_2/h_1 by η, the thickness h at any x can be expressed as

$$h = \frac{2h_o}{(\eta + 1)}\left[1 + (\eta - 1)\frac{x}{a}\right]$$

Hence,

$$D = \frac{E}{12(1 - \mu^2)}\frac{8h_o^3}{(\eta + 1)^3}\left[1 + (\eta - 1)\frac{x}{a}\right]^3$$

and

$$\bar{\rho} = \rho h = \frac{2\rho h_o}{(\eta + 1)}\left[1 + (\eta - 1)\frac{x}{a}\right]$$

where ρ is the mass density.

In view of the symmetry of the geometry, the boundary conditions and the assumed mode shape about both the x and y axes, it is sufficient to calculate energies for one quarter of the plate. This calculation yields the following:

$$U_e = \frac{Eh_o^3\psi A^2}{24a^2(1 - \mu^2)}\left[\frac{2\pi^4}{\psi} - 24\pi^2 + \pi^2(9 + 2\pi^2)\psi - 24(\pi^2 - 8)\psi^2\right]$$

$$T = \frac{3\rho a^2 h_o \dot{A}^2}{8\pi^2}(3\pi^2 - 16\psi), \quad \text{with } \psi = \frac{\eta - 1}{\eta + 1}$$

Hence,

$$\omega_{\text{tap.}}^2 = \frac{\pi^2 Eh_o^2\psi}{9\rho a^4(1 - \mu^2)}\left[\frac{\frac{2\pi^4}{\psi} - 24\pi^2 + \pi^2(9 + 2\pi^2)\psi - 24(\pi^2 - 8)\psi^2}{3\pi^2 - 16\psi}\right]$$

Specializing the above for the case of the uniform plate by taking $\eta = 1$, one gets

$$\omega_{\text{uni.}}^2 = \frac{2\pi^4 Eh_o^2}{27\rho a^4(1 - \mu^2)}$$

Thus, finally, $(\omega_{\text{tap.}}^2/\omega_{\text{uni.}}^2)$ can be expressed in terms of η, and hence its dependence on η can be studied easily by making a plot.

(At this stage, one should ponder for a while on how the influence of η would have been studied had one adopted a numerical approach like the finite difference method or the finite element method, and appreciate the superiority of the analytical approach. It is for this reason that analytical methods are often preferred to numerical methods for optimization studies).

Since η always occurs in the form of ψ in the final result, it is convenient to make a plot with respect to ψ. The extreme values of η, corresponding to the tapered configurations shown in Fig. 8.8, are 0 and ∞. The corresponding values of ψ (obtained, for the case of $\eta = \infty$, by evaluating the limit) are -1 and $+1$, respectively. The plot of $\left(\omega^2_{tap.} / \omega^2_{uni.} \right)$ versus ψ is shown in Fig. 8.9.

It is clear that a significant change in the fundamental frequency can be achieved by employing a taper. The best configuration, given by $\psi = -1$ (i.e. $\eta = 0$), is shown in Fig. 8.8a. (It should be pointed out that this conclusion is purely from the viewpoint of a high natural frequency; the suggested shape is clearly impractical because the stresses at the clamped edges would be infinitely large.)

Fig. 8.8 Extreme cases

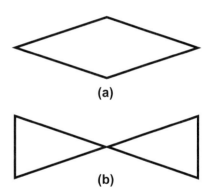

(a)

(b)

Fig. 8.9 Influence of the taper parameter

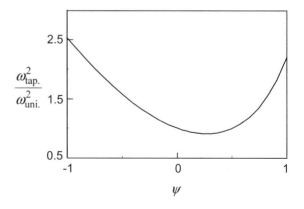

8.3 Galerkin's Method

Galerkin's method, as compared to Rayleigh–Ritz method, belongs to a different category in that it deals with the differential equation directly instead of the energy functional. Thus, it is a more general method because there are problems which cannot be posed in terms of minimization of energy or some such functional.

Galerkin's method is a special case of what is known as the *method of weighted residuals*. Considering a plate, and writing the governing equation in operator form as

$$L(w) - q = 0 \tag{8.9}$$

where L is a differential operator, one seeks, as in Rayleigh–Ritz method, an approximate solution in terms of known space functions ϕ_i as

$$w = \sum_{i=1}^{N} A_i \phi_i \tag{8.10}$$

Here, $\phi_i(x, y)$ are functions that satisfy both the kinematic and the natural boundary conditions and are often referred to as *comparison functions* to distinguish them from the admissible functions of Rayleigh–Ritz method. Substitution for w in the governing equation yields an error, called the residual R, given by

$$R = L\left(\sum A_i \phi_i\right) - q \tag{8.11}$$

In the method of weighted residuals, one tries to minimize this residual by orthogonalizing it over the domain with respect to a set of N suitably chosen weighting functions $\psi_i(x, y)$. This means that

$$\int R\psi_i \, d\text{Area} = 0, \quad i = 1, 2, \ldots, N \tag{8.12}$$

which are N algebraic equations in terms of the constants A_i. These equations can be solved to obtain the constants and hence the approximate deflection function.

In Galerkin's method, the weighting functions $\psi_i(x, y)$ are taken to be the same as the approximating functions $\phi_i(x, y)$. Hence, the final algebraic equations to determine A_i are given by

$$\int R\phi_i \, d\text{Area} = 0, \quad i = 1, 2, \ldots, N \tag{8.13}$$

where the integral on the left-hand side is known as Galerkin's integral.

It should be clear from the above exposition that Galerkin's method is applicable to any problem with a known governing equation, and hence to non-conservative (i.e. dissipative) systems as well, for which Rayleigh–Ritz method is not applicable because the strain energy functional is not defined.

Another difference between the two methods is that *all* the boundary conditions have to be satisfied while choosing the approximating functions in Galerkin's method while only kinematic boundary conditions need be satisfied in Rayleigh–Ritz method. It can be proved mathematically that results by both these methods would be identical for a problem if one employs the same functions satisfying all the boundary conditions.

The application of Galerkin's method comprises of the choice of the functions ϕ_i and the evaluation of Galerkin's integral; the methodology is the same irrespective of whether the analysis is for static bending or buckling or vibrations. This can be seen from the following examples.

Example 8.8 Flexure of a simply supported square plate under transverse sinusoidal load.

The plate $(0 \leq x, y \leq a)$, simply supported all around, undergoes bending due to the load given by

$$q = q_o \sin \frac{\pi x}{a} \sin \frac{\pi y}{a}$$

This problem was solved earlier by Rayleigh–Ritz method using admissible functions satisfying the kinematic boundary conditions alone (Example 8.1). Here, one needs to satisfy all the boundary conditions. Such comparison functions can be derived formally as follows.

Choosing the approximate solution in variables-separable form as

$$w(x, y) = A X(x) Y(y)$$

$X(x)$ should be such that

$$X(0) = X(a) = X_{,xx}(0) = X_{,xx}(a) = 0.$$

Starting with a general polynomial for X as

$$X = k_1 + k_2 x + k_3 x^2 + k_4 x^3 + x^4$$

the constants k_1 to k_4 can be determined using the above four conditions. Proceeding in this manner, one gets, after factorization,

$$X = x(x - a)(x^2 - ax - a^2)$$

Similarly, $Y = y(y - a)(y^2 - ay - a^2)$.
Hence, $w = Ax(x - a)(x^2 - ax - a^2)y(y - a)(y^2 - ay - a^2)$.

(If one needs more such functions for use in a multi-term solution of the form

$$w = \sum_m \sum_n A_{mn} X_m Y_n,$$

one should seek X_m in a more general form as

$$X_m = (k_{1m} + k_{2m}x + k_{3m}x^2 + k_{4m}x^3 + x^4)x^{m-1}, \quad m = 1, 2, \ldots,$$

and find the constants in terms of m, and similarly for Y_n.)

Substitution for w in the governing equation given by

$$D\nabla^4 w = q_o \sin \frac{\pi x}{a} \sin \frac{\pi y}{a}$$

yields the residual as

$$R = D\nabla^4[Ax(x-a)(x^2 - ax - a^2)y(y-a)(y^2 - ay - a^2)]$$
$$- q_o \sin \tfrac{\pi x}{a} \sin \tfrac{\pi y}{a}$$

The next step is to obtain Galerkin's integral and equate it to zero,

i.e. $$\int_0^a \int_0^a Rx(x-a)(x^2 - ax - a^2)y(y-a)(y^2 - ay - a^2)\, dx\, dy = 0$$

which, after simplification, gives

$$\frac{694}{735} Da^{14} A - \frac{2304}{\pi^{10}} a^{10} q_o = 0$$

i.e. $$A = 0.02606 \frac{q_0}{a^4 D}$$

This yields

$$w_{\max} = 0.002545 \frac{q_o a^4}{D} \quad (-0.86\% \text{ error})$$

$$M_{x\ \max} = 0.0317 q_o a^2 \quad (-3.56\% \text{ error})$$

As mentioned earlier, if one uses the above approximating function with Rayleigh–Ritz method, one would have obtained the same final results (see Problem 5). Thus, in this problem, the satisfaction of the natural boundary conditions besides the essential conditions results in better accuracy, as can be seen by comparing the above results with those of Example 8.1. (However, it should be noted that this is not true for all problems and sometimes Rayleigh–Ritz method may yield worse results with an approximating function satisfying all the boundary conditions than with one satisfying the essential conditions alone.)

Example 8.9 Buckling of a simply supported square plate due to uniaxial compression.

Let the plate ($0 \leq x, y \leq a$) be subjected to uniform compression in the x-direction ($N_x = -P_{cr}$). The admissible function used in Example 8.8 is suitable for evaluating the lowest buckling load.

Galerkin's methodology leads to the following mathematical steps:

$$R = D\nabla^4\big[Ax(x-a)(x^2-ax-a^2)y(y-a)(y^2-ay-a^2)\big]$$
$$+P_{cr}\big[Ax(x-a)(x^2-ax-a^2)y(y-a)(y^2-ay-a^2)\big]_{xx}$$

$$\iint Rx(x-a)(x^2-ax-a^2)y(y-a)(y^2-ay-a^2)dx\,dy = 0$$

which becomes, after simplification,

$$\frac{694}{735}Da^{14}A - \frac{527}{22,050}a^{16}P_{cr}A = 0$$

Hence, for a non-trivial solution, $P_{cr} = 39.51\frac{D}{a^2}$, which is very close to the exact solution (see Example 7.1) given by

$$P_{cr} = 4\pi^2\frac{D}{a^2} \approx 39.48\frac{D}{a^2}$$

One should again note that the approximate solution is an upper bound.

Example 8.10 Fundamental frequency of a simply supported square plate.

The plate considered in the preceding two examples is now analysed for its fundamental frequency. Continuing with the same admissible function, which, for the present time-dependent problem, becomes

$$w = A(t)x(x-a)(x^2-ax-a^2)y(y-a)(y^2-ay-a^2),$$

the residual is given by

$$R = D\nabla^4\big[A(t)x(x-a)(x^2-ax-a^2)y(y-a)(y^2-ay-a^2)\big]$$
$$+\bar{\rho}\ddot{A}x(x-a)(x^2-ax-a^2)y(y-a)(y^2-ay-a^2)$$

Then, we have

$$\int_0^a\int_0^a Rx(x-a)(x^2-ax-a^2)y(y-a)(y^2-ay-a^2)dxdy = 0$$

which yields

$$\frac{961}{396,900}\overline{\rho}a^{18}\ddot{A} + \frac{694}{735}Da^{14}A = 0$$

So, the natural frequency is given by

$$\omega^2 = \frac{694 \times 396,900}{735 \times 961}\frac{D}{\overline{\rho}a^4} = 389.97\frac{D}{\overline{\rho}a^4}$$

which agrees very closely (once again, from the higher side) with the exact value (see Example 6.1) of $\frac{4\pi^4 D}{\overline{\rho}a^4}$.

Example 8.11 Fundamental frequency and mode shape of a uniaxially compressed SCSC plate.

Let us consider the free vibration of a rectangular plate ($0 \le x, y \le a, b$) with $a = 5b$ and the edges $y = 0, b$ clamped and the others simply supported and subjected to uniform compression $N_x = -P$. This is similar to the problem of Example 7.4 and will now be solved using Galerkin's method, with due care to consider mode shapes with a generic number of nodal lines parallel to the y-axis. The fundamental mode shape is not expected to have any nodal lines parallel to x-axis.

A convenient comparison function is

$$w = A(t) \sin \frac{m\pi x}{a}y^2(y - b)^2$$

Putting this in the governing equation

$$D\nabla^4 w + \overline{\rho}w_{,tt} = N_x w_{,xx}$$

and minimizing the residual using Galerkin's integral, one gets

$$A_{,tt} + A\left[\frac{D(m^4\pi^4 + 600m^2\pi^2 + 315,000)}{625\overline{\rho}b^4} - \frac{Pm^2\pi^2}{25\overline{\rho}b^2}\right] = 0$$

Thus the natural frequency corresponding to m half-waves in the x-direction is given by

$$\omega^2 = \left[\frac{D(m^4\pi^4 + 600m^2\pi^2 + 315,000)}{625\overline{\rho}b^4} - \frac{Pm^2\pi^2}{25\overline{\rho}b^2}\right]$$

Noting that this should be equated to zero to get the critical values of P corresponding to buckling of the plate, and seeking the value of the lowest P_{cr} by trying out different values of m, one can easily obtain

$$P_{cr} = \left[\frac{D(m^4\pi^4 + 600m^2\pi^2 + 315,000)}{25m^2\pi^2 b^2}\right]_{\text{lowest}}$$

$$= 69.21\frac{D}{b^2} \text{ for } m = 8$$

Just as in Example 7.4, let us normalize the compressive force P with this critical value as $P = \eta P_{cr}$ so that the expression for the natural frequency can be rewritten as

$$\omega^2 = \frac{D}{\bar{\rho}b^4}\left[\frac{(m^4\pi^4 + 600m^2\pi^2 + 315{,}000)}{625} - 2.769\eta m^2\pi^2\right]$$

The fundamental frequency can now be obtained by seeking the lowest value of the above, for any specified η, by trying out different values of m, the extreme values of $\frac{\bar{\rho}b^4\omega^2}{D}$ being 513.6 ($m = 1$) for $\eta = 0$, and zero ($m = 8$) for $\eta = 1$. Thus, the fundamental mode shape changes as the compressive force is varied; for example, $m = 2$ for $\eta = 0.4$ and $m = 5$ for $\eta = 0.6$, with the frequency parameter being 500.7 and 428.4, respectively. Though the values of the frequencies are approximate and may require a multi-term Galerkin solution (with more terms in the y-direction) for better accuracy, the fundamental mode shapes as identified by this one-term solution are expected to be correct.

Summary

After a brief introduction to approximate methods in general, two commonly used approximate analytical methods—those of Rayleigh–Ritz and Galerkin—have been explained and applied to the study of bending, buckling and free vibrations of rectangular plates. It has been clearly pointed out that these approximations represent a stiffer model than the original structure, and hence yield upper bounds of the natural frequencies and the buckling loads.

It is appropriate to state here that these methods are used, often only with the single-term approximation, primarily to obtain a first estimate of the deflection or frequency or buckling load. Though one can obtain convergence to the exact solution by taking a large number of terms chosen according to certain (known and well-defined) criteria, such studies are not common; numerical methods, which can be easily carried out by the use of computers, are preferred in such cases. Further, there is no alternative to the numerical approach for complicated problems such as those involving irregular shapes, or point supports, or edges where the boundary conditions are not the same over the entire length.

Conceptual Questions

1. From a book dealing with variational principles for structural mechanics, find out regarding:
(a) the principle of virtual work, and how the principle of minimum potential energy is derived from it;
(b) the principle of complementary virtual work;
(c) the principle of complementary energy;

(d) mixed principles such as those of Hellinger-Reissner and Hu-Washizu;

(e) Hamilton's principle, and how it is related to Lagrange's equations of motion;

(f) Rayleigh's method for determining the fundamental frequency; Rayleigh's quotient; and Rayleigh–Ritz method looked upon as the technique of minimizing Rayleigh's quotient;

(g) the mathematical equivalence of Rayleigh–Ritz and Galerkin's methods when the natural boundary conditions are also satisfied.

2. Collocation is another popular approximate analytical method besides Rayleigh–Ritz and Galerkin methods. Refer to a book dealing with approximate solutions and get to know the principle behind this method.

3. Indicate how Galerkin's method would be applied for analysing a plate with a concentrated load.

Problems
Section 8.2

1. Choose suitable admissible functions, in terms of polynomials, for use with Rayleigh–Ritz method for (a) SCSC; (b) SFCC; and (c) CFSS plates. Do this such that the number of nodal lines parallel to either of x and y axes is (i) 0; (ii) 1; and (iii) 2.

2. Repeat Problem 1 using trigonometric functions.

3. A simply supported infinite strip undergoes cylindrical bending due to a centrally applied line load as shown. Express the potential energy in terms of w for this case. Using this expression with a suitable one-term approximation for w, determine the maximum deflection by Rayleigh–Ritz method. Compare the result with the exact value. (*Hint: See Question 2 under* **Conceptual Questions** *in Chap. 2.*)

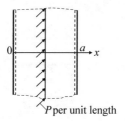

P per unit length

Problem 3

4. For a plate clamped at all edges, show that, irrespective of the choice of the admissible function, the part of the strain energy proportional to $\iint (w_{,xx}w_{,yy} - w_{,xy}^2)\, dx\, dy$ is identically zero. Do this mathematically by using the clamped edge conditions and integration by parts.

5. With reference to Example 8.8, reproduce the final results using Rayleigh–Ritz method.

6. Using a one-term Rayleigh–Ritz method, obtain the maximum slope in the y-direction for the plates shown.

7. The clamped square plate shown is subjected to a moment M applied at the centre O. Obtain the deflection at $(a/2, -a/2)$ using a suitable one-term Rayleigh–Ritz method. (*Hint: Resolve M along the two axes, solve the problem for either of the components, and finally use superposition to obtain the solution.*)

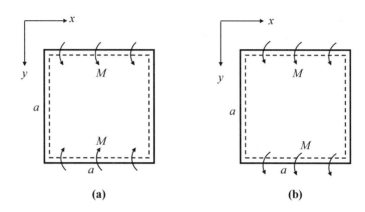

(a)   (b)

Problem 6

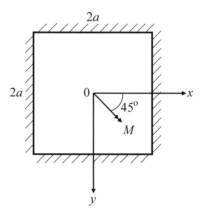

Problem 7

8. A simply supported rectangular plate is subjected to four concentrated transverse loads, P each, acting at the centres of the four quadrants as shown. Using a one-term Rayleigh–Ritz solution, determine the deflection directly under a load. Plot the deflection along a diagonal, taking care to show the slope at the centre correctly.

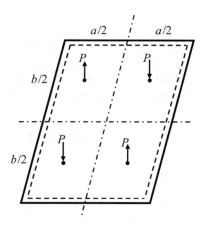

Problem 8

9. With reference to Example 8.5, considering a square plate, plot the buckled shapes along the two diagonals for the two critical loads obtained. From these plots, it should be inferred that a higher buckling load always corresponds to a case of more severe deformation (i.e. change of shape). One should also notice that the diagonals are lines of symmetry or antisymmetry—explain why they should be so.

10. A square plate of side a, simply supported at all the four edges, has a central, symmetrically located square hole of side $a/3$. Using one-term Rayleigh–Ritz method, determine (a) the buckling load corresponding to equal biaxial compression; (b) the fundamental frequency. Compare the results with those for a plate without the hole. Take $\mu = 0.3$.

11. A simply supported square plate ($0 \leq x, y \leq a$) has a thickness h in the region $0 \leq x \leq a/2$, and a thickness $2h$ in the remaining half. Obtain the natural frequencies of the plate corresponding to the two-term Rayleigh–Ritz approximation given by

$$w = A(t) \sin \frac{\pi x}{a} \sin \frac{\pi y}{a} + B(t) \sin \frac{2\pi x}{a} \sin \frac{\pi y}{a}$$

Plot the x-variation of the two mode shapes. Take $\mu = 0.3$.

12. Using one-term Rayleigh–Ritz method, obtain the fundamental frequency of a clamped square plate. If a concentrated mass equal to one-third of the mass of the plate were to be lumped at the centre, find out its effect on the fundamental frequency.

Section 8.3

13. A simply supported square plate is subjected to a transverse line load of intensity P per unit length as shown. Obtain the maximum deflection using one-term Rayleigh–Ritz method with an approximating function satisfying all the boundary conditions. Reproduce the same result by employing Galerkin's method.

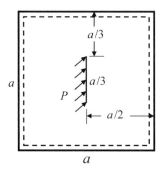

Problem 13

14. A CCCC square plate $(0 \leq x, y \leq a)$ is subjected to a known uniform in-plane pressure p acting along the edges $x = 0$ and $x = a$. Using the solution given by Galerkin's method with the admissible function

$$w = A\left(1 - \cos\frac{2\pi x}{a}\right)\left(1 - \cos\frac{2\pi y}{a}\right)$$

determine the critical value of an additional compressive force that causes buckling when applied along $y = 0$ and $y = a$.

15. (a) An infinite strip $(0 \leq x \leq a; -\infty \leq y \leq \infty)$ of thickness h_o, simply supported at $x = 0$ and clamped at $x = a$, is subjected to a uniform compressive force applied at these edges. Show that

$$w = A\left(a^3 x - 3ax^3 + 2x^4\right)$$

can be used as a trial function for determining the buckling load by Galerkin's method. Find the value of this load and compare it with the exact solution. (*Hint: See Question* 2 *under* **Conceptual Questions** *in Chap.* 2)

(b) Suppose one wishes to increase the buckling load of the strip without increasing the weight, by using a tapered cross section as shown. Assuming the one-term approach of (a) to be acceptable, show that the optimum value of c is given by $c = 0.392a$. Compare the resulting buckling load with that of (a) and that corresponding to $c = a/2$.

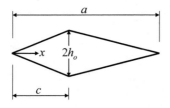

Problem 15 (b)

16. Using a suitable one-term Galerkin's method, find out the frequency corresponding to cylindrical bending of an infinite plate strip clamped at one longitudinal edge and free at the other.

17. The exact solution of a circular plate subjected to uniform load on a concentric circular area was asked for in Problem 9 of Chap. 5. Redo this problem by using Galerkin's method with a suitable one-term approximation, and compare the result with the exact solution.

Part II
Complicating Effects and Corresponding Theories

Introduction to Part II

The introductory content of Part I has to be supplemented by theories accounting for various complicating factors such as material anisotropy, discretely or continuously varying material properties, variable thickness, and nonlinear deformations, because these are very commonly encountered in modern plate structures. While such theories can still be based on Kirchhoff-Poisson hypothesis, this assumption itself needs to be discarded when the plate can no more be classified as thin; in that case, one can use a less restrictive assumption to develop a refined plate theory or carry out three-dimensional analysis using the theory of elasticity.

The objective of this second part of the book is to familiarize the reader with such complications, and to provide appropriate references for further study. As compared to the lucid explanatory style of Part I, the presentation here is quite concise and the reader is expected to put in correspondingly more effort. The emphasis is on highlighting the influence of each complicating factor on the behavior of the plate, and towards this goal, attention is confined to fairly simple examples under each head. No conceptual questions or exercise problems are included here, and anyone interested in pursuing a topic further would do well to study the suggested references and their cross-references diligently and to reproduce some results obtained therein.

Chapter 9
Anisotropic, Laminated and Functionally Graded Plates

The purpose of this chapter is to modify the theory presented in Chap. 2 to account for a change of the constituent material - from the simple homogeneous isotropic metallic material to one that is anisotropic, or heterogeneous with multiple isotropic or fibre composite layers of different properties, or one that has its properties varying through the thickness in a continuous fashion—the so-called functionally graded material. Of these modifications, the first is the easiest to handle because it does not require a change in the displacement field, while the other two require changes in view of a possible coupling of the bending and stretching modes. It should be noted that the resulting theories should all be termed Classical, since they are all based on Kirchhoff–Poisson hypothesis; the range of applicability of these theories will be discussed in the next chapter.

9.1   CPT for Homogeneous Anisotropic Plates

9.1.1   *The Anisotropic Constitutive Law*

The anisotropy we seek to account for is best understood by first considering the simpler case of *orthotropy*. A three-dimensional orthotropic body is one for which the stress–strain law is of the following form.

$$
\begin{pmatrix} \sigma_1 \\ \sigma_2 \\ \sigma_3 \end{pmatrix} = \begin{bmatrix} C_{11} & C_{12} & C_{13} \\ & C_{22} & C_{23} \\ \text{sym.} & & C_{33} \end{bmatrix} \begin{pmatrix} \varepsilon_1 \\ \varepsilon_2 \\ \varepsilon_3 \end{pmatrix}
$$

$$
\tau_{23} = C_{44}\gamma_{23}; \quad \tau_{13} = C_{55}\gamma_{13}; \quad \tau_{12} = C_{66}\gamma_{12} \tag{9.1}
$$

© The Author(s) 2021
K. Bhaskar and T. K. Varadan, *Plates*,
https://doi.org/10.1007/978-3-030-69424-1_9

with respect to the special set of mutually orthogonal axes 1-2-3 referred to as the principal axes of orthotropy or simply material axes. C_{ij} are the three-dimensional stiffness coefficients. The behaviour of this body, when studied with respect to the 1-2-3 axes, is similar to that of an isotropic body in the sense there is no coupling between stretch and shear modes or between the different shear modes; the only difference between the two is that the stiffness coefficients are now different in different directions and there are nine independent coefficients here as compared to just two for the isotropic body (see Eq. (1.3)).

In order to understand this better, let us define Young's moduli E_1, E_2 and E_3 along the 3-directions and shear moduli G_{23}, G_{13} and G_{12} in the three corresponding planes. Let us also define Poisson's ratios. Because of orthotropy one needs to define two Poisson's ratios corresponding to any set of two axes; for example, corresponding to 1-2 axes, one has

$$\mu_{12} = -\varepsilon_2/\varepsilon_1 \text{ when } \sigma_1 \text{ alone is applied}$$
$$\mu_{21} = -\varepsilon_1/\varepsilon_2 \text{ when } \sigma_2 \text{ alone is applied} \tag{9.2}$$

and these are not equal. Similarly, one can define the ratios $\mu_{23}, \mu_{32}, \mu_{13}$ and μ_{31}.

In terms of the above elastic constants, one can now express the strain components in terms of the stress components for any general loading as

$$
\begin{array}{ll}
\varepsilon_1 = \frac{\sigma_1}{E_1} - \frac{\mu_{21}\sigma_2}{E_2} - \frac{\mu_{31}\sigma_3}{E_3} & \gamma_{23} = \frac{\tau_{23}}{G_{23}} \\
\varepsilon_2 = -\frac{\mu_{12}\sigma_1}{E_1} + \frac{\sigma_2}{E_2} - \frac{\mu_{32}\sigma_3}{E_3} & \gamma_{13} = \frac{\tau_{13}}{G_{13}} \\
\varepsilon_3 = -\frac{\mu_{13}\sigma_1}{E_1} - \frac{\mu_{23}\sigma_2}{E_2} + \frac{\sigma_3}{E_3} & \gamma_{12} = \frac{\tau_{12}}{G_{12}}
\end{array}
\tag{9.3}
$$

By virtue of Maxwell's reciprocal theorem, the strain ε_1 due to unit stress σ_2 alone should be equal to the strain ε_2 due to unit stress σ_1 alone and so on, leading to the following relations

$$\frac{\mu_{12}}{E_1} = \frac{\mu_{21}}{E_2}; \quad \frac{\mu_{13}}{E_1} = \frac{\mu_{31}}{E_3}; \quad \frac{\mu_{23}}{E_2} = \frac{\mu_{32}}{E_3} \tag{9.4}$$

Thus, for the orthotropic body, there are nine independent elastic constants—the three E's, the three G's and three μ's, say μ_{12}, μ_{23} and μ_{13}. This is in contrast to only one E and one μ for the isotropic body valid for all directions and only one G dependent on them, i.e. just two independent elastic constants.

Let us now consider a thin orthotropic plate—thin in the 3-direction. This would require a plane-stress reduction exactly as was done earlier for the isotropic case (see Eq. (2.9)), leading to these final equations:

$$
\begin{aligned}
\sigma_1 &= \frac{E_1}{(1 - \mu_{12}\mu_{21})}(\varepsilon_1 + \mu_{21}\varepsilon_2) \\
\sigma_2 &= \frac{E_2}{(1 - \mu_{12}\mu_{21})}(\varepsilon_2 + \mu_{12}\varepsilon_1) \\
\tau_{12} &= G_{12}\gamma_{12}
\end{aligned}
\tag{9.5}
$$

which can be rewritten in terms of the appropriately defined plane-stress reduced stiffness coefficients Q_{ij} as

$$
\begin{pmatrix} \sigma_1 \\ \sigma_2 \\ \tau_{12} \end{pmatrix} = \begin{bmatrix} Q_{11} & Q_{12} & 0 \\ & Q_{22} & 0 \\ \text{sym.} & & Q_{66} \end{bmatrix} \begin{pmatrix} \varepsilon_1 \\ \varepsilon_2 \\ \gamma_{12} \end{pmatrix}
\tag{9.6a}
$$

or in shortened form,

$$
\{\sigma\}_{1-2} = [Q]\{\varepsilon\}_{1-2}
\tag{9.6b}
$$

Examples of orthotropic plates include those made of orthotropic materials like wood (for which stiffnesses along and across the grains are different) or those that display orthotropy due to stiffening or unidirectional fibre reinforcement (see Fig. 9.1).

Let us now consider the case of a unidirectional fibre-reinforced composite rectangular plate wherein the fibres are not parallel to the edges as in Fig. 9.1, but to an inclined in-plane direction as shown in Fig. 9.2. If such a plate is subjected to simple tension along x or y direction, it is easy to imagine that the fibres would tend to align themselves in the direction of loading, thus leading to a shear strain γ_{xy} as shown. Thus, with respect to x-y coordinates, there is a coupling between stretch and shear modes—which also manifests itself as linear strains ε_x and ε_y when subjected to pure shear τ_{xy}—while such coupling would be absent if one uses the fibre direction 1 and the in-plane transverse direction 2 as the reference axes.

While the stretch–shear coupling described above is for an orthotropic plate with reference to inclined axes different from the principal material directions, this kind

Fig. 9.1 a Stiffened plate, **b** fibre-reinforced plate

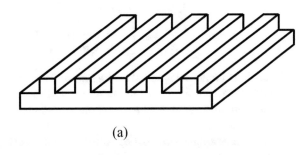

(a)

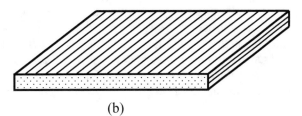

(b)

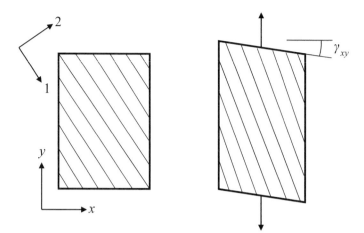

Fig. 9.2 Stretch–shear coupling

of behaviour is also seen for plates made of some naturally occurring materials with respect to every set of in-plane axes. We shall confine ourselves, in this book, to such behaviour characterized by coupling of in-plane shear with the stretch modes but with these modes totally uncoupled with respect to the transverse shear modes.

In other words, the kind of anisotropy we would like to address in this book is described by the following three-dimensional constitutive law.

$$\begin{pmatrix} \sigma_x \\ \sigma_y \\ \sigma_z \\ \tau_{xy} \end{pmatrix} = \begin{bmatrix} C_{11} & C_{12} & C_{13} & C_{16} \\ & C_{22} & C_{23} & C_{26} \\ & & C_{33} & C_{36} \\ \text{sym.} & & & C_{66} \end{bmatrix} \begin{pmatrix} \varepsilon_x \\ \varepsilon_y \\ \varepsilon_z \\ \gamma_{xy} \end{pmatrix}$$

$$\begin{pmatrix} \tau_{yz} \\ \tau_{xz} \end{pmatrix} = \begin{bmatrix} C_{44} & C_{45} \\ \text{sym.} & C_{55} \end{bmatrix} \begin{pmatrix} \gamma_{yz} \\ \gamma_{xz} \end{pmatrix} \tag{9.7}$$

This kind of material is called *monoclinic* and is a special case of the most general anisotropic material for which all the stretch and shear modes are coupled with one another. Note that the monoclinic material is also characterized by coupling of the two transverse shear modes.

A simple example will suffice to differentiate between monoclinic and fully anisotropic behaviour; consider a multi-directionally reinforced fibrous composite with the following two possibilities: (a) parallel fibres run in two non-orthogonal directions with both of them being perpendicular to the vertical direction; (b) parallel fibres run in three non-orthogonal directions (Fig. 9.3). The first of these is a mono-clinic material; here it is easy to visualize the absence of distortion of the vertical faces from their original rectangular shapes into parallelograms, when the body is subjected to stretching, say, in the vertical direction, though the horizontal faces

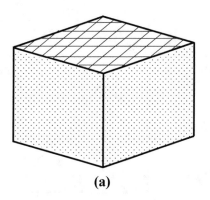 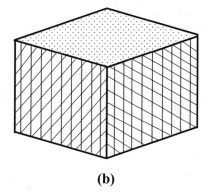

(a)   (b)

Fig. 9.3 **a** Monoclinic material, **b** anisotropic material

undergo distortion. A similar loading on the second material produces distortion of all the faces and this is characteristic of general anisotropy.

For a thin plate made of a monoclinic material, the plane-stress reduced constitutive law is given by

$$
\begin{pmatrix} \sigma_x \\ \sigma_y \\ \tau_{xy} \end{pmatrix} = \begin{bmatrix} \overline{Q}_{11} & \overline{Q}_{12} & \overline{Q}_{16} \\ & \overline{Q}_{22} & \overline{Q}_{26} \\ \text{sym.} & & \overline{Q}_{66} \end{bmatrix} \begin{pmatrix} \varepsilon_x \\ \varepsilon_y \\ \gamma_{xy} \end{pmatrix}
\tag{9.8a}
$$

or in shortened form,

$$
\{\sigma\}_{x-y} = [\overline{Q}]\{\varepsilon\}_{x-y}
\tag{9.8b}
$$

This is obtained by eliminating ε_z from the three-dimensional monoclinic constitutive law using $\sigma_z = 0$. This law is also valid for the orthotropic plate when referred to x-y axes which are inclined with respect to the material axes 1-2 as explained earlier with reference to Fig. 9.2. In this case, if the orthotropic elastic constants E_1, E_2, μ_{12} and G_{12} are known from appropriate material characterization tests, then one can calculate the plane-stress reduced stiffness coefficients Q_{11}, Q_{12}, Q_{22} and Q_{66} with reference to the material axes (see Eqs. (9.5, 9.6a)) and thereafter determine the coefficients $\overline{Q}_{ij}$ using the angle of inclination between the two sets of reference axes. For this reason, $\overline{Q}_{ij}$ are referred to as the transformed plane-stress reduced stiffness coefficients.

Carrying out a 2-D transformation from x-y axes to 1-2 axes by considering the equilibrium of a small triangular element (Fig. 9.4), one can easily derive the relations between the corresponding stress components. These can be expressed as

Fig. 9.4 Transformation
from x-y to 1-2 axes

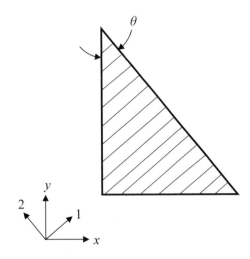

$$\begin{pmatrix} \sigma_1 \\ \sigma_2 \\ \tau_{12} \end{pmatrix} = \begin{bmatrix} c^2 & s^2 & 2cs \\ s^2 & c^2 & -2cs \\ -cs & cs & c^2 - s^2 \end{bmatrix} \begin{pmatrix} \sigma_x \\ \sigma_y \\ \tau_{xy} \end{pmatrix} \tag{9.9}$$

where $c = \cos\theta$, $s = \sin\theta$. Noting that the stress components and the strain components follow the same transformation rules because they are both second-order tensor components, and that the tensorial shear strain ε_{ij} defined as $\gamma_{ij}/2$ has to be used in that context, one can immediately write

$$\begin{pmatrix} \varepsilon_1 \\ \varepsilon_2 \\ \varepsilon_{12} \end{pmatrix} = \begin{bmatrix} c^2 & s^2 & 2cs \\ s^2 & c^2 & -2cs \\ -cs & cs & c^2 - s^2 \end{bmatrix} \begin{pmatrix} \varepsilon_x \\ \varepsilon_y \\ \varepsilon_{xy} \end{pmatrix}$$

and hence,

$$\begin{pmatrix} \varepsilon_1 \\ \varepsilon_2 \\ \gamma_{12} \end{pmatrix} = \begin{bmatrix} c^2 & s^2 & cs \\ s^2 & c^2 & -cs \\ -2cs & 2cs & c^2 - s^2 \end{bmatrix} \begin{pmatrix} \varepsilon_x \\ \varepsilon_y \\ \gamma_{xy} \end{pmatrix} \tag{9.10}$$

Use of Eqs. (9.9, 9.10) in Eq. (9.6a) and rearrangement yields the constitutive relations with reference to x-y coordinates. Denoting the transformation matrices in Eqs. (9.9) and (9.10) as $[T]$ and $[T_1]$, respectively, one can rewrite Eq. (9.6b) as

$$[T]\{\sigma\}_{x-y} = [Q][T_1]\{\varepsilon\}_{x-y}$$

i.e. $\{\sigma\}_{x-y} = [T]^{-1}[Q][T_1]\{\varepsilon\}_{x-y}$

Thus, one has

$$[\overline{Q}] = [T]^{-1}[Q][T_1]$$

which leads to the following relations:

$$\overline{Q}_{11} = c^4 Q_{11} + s^4 Q_{22} + 2c^2 s^2 (Q_{12} + 2Q_{66})$$
$$\overline{Q}_{22} = s^4 Q_{11} + c^4 Q_{22} + 2c^2 s^2 (Q_{12} + 2Q_{66})$$
$$\overline{Q}_{12} = c^2 s^2 (Q_{11} + Q_{22} - 4Q_{66}) + (c^4 + s^4) Q_{12}$$
$$\overline{Q}_{66} = c^2 s^2 (Q_{11} + Q_{22} - 2Q_{12}) + (c^2 - s^2)^2 Q_{66}$$
$$\overline{Q}_{16} = cs[c^2 Q_{11} - s^2 Q_{22} - (c^2 - s^2)(Q_{12} + 2Q_{66})]$$
$$\overline{Q}_{26} = cs[s^2 Q_{11} - c^2 Q_{22} + (c^2 - s^2)(Q_{12} + 2Q_{66})] \tag{9.11}$$

Here, it should be noted that $\overline{Q}_{16}$ and $\overline{Q}_{26}$ are odd functions of θ while the others are even functions of θ.

For the special case of $\theta = 90°$, one gets

$$\overline{Q}_{11} = Q_{22}; \ \overline{Q}_{22} = Q_{11}; \ \overline{Q}_{12} = Q_{12}; \ \overline{Q}_{66} = Q_{66}; \ \overline{Q}_{16} = \overline{Q}_{26} = 0 \tag{9.12}$$

When the material axes 1-2 coincide with the x-y axes with $\theta = 0°$ or $90°$, the plate is referred to as *specially orthotropic*; for other orientations, the plate is referred to as *generally orthotropic*, or sometimes simply *anisotropic*.

Before proceeding on to the analysis of these plates, it is appropriate to briefly discuss the determination of the orthotropic elastic constants E_1, E_2, G_{12} and μ_{12} for the commonly encountered case of a unidirectional fibre composite. As mentioned earlier, these can be obtained by appropriate material characterization tests; alternatively, they can also be calculated using some mathematical relations if the material properties of the constituent matrix and fibre phases are known along with the fibre-packing geometry. Among several such *micromechanics* equations,[1] the simplest ones, referred to as the *rule of mixtures*, are the most commonly used. They are given by

$$E_1 = E_f V_f + E_m (1 - V_f)$$
$$\mu_{12} = \mu_f V_f + \mu_m (1 - V_f)$$
$$\frac{1}{E_2} = \frac{V_f}{E_f} + \frac{(1 - V_f)}{E_m}$$
$$\frac{1}{G_{12}} = \frac{V_f}{G_f} + \frac{(1 - V_f)}{G_m} \tag{9.13}$$

---

[1] R. F. Gibson, Principles of Composite Material Mechanics, CRC Press, 2007.

where the subscripts m and f refer to the matrix and fibre phases, respectively, and V_f is the volume fraction of the fibre reinforcement defined as the volume of fibres per unit volume of the composite material.

9.1.2 Plate Equations

The development of the classical theory for anisotropic plates is on the same lines as for the isotropic plate, but for the change in the constitutive law.

The moment–curvature relations are obtained as

$$
\begin{pmatrix} M_x \\ M_y \\ M_{xy} \end{pmatrix} = \int_{-h/2}^{h/2} \begin{pmatrix} \sigma_x \\ \sigma_y \\ \tau_{xy} \end{pmatrix} z\, dz
$$

$$
= \int_{-h/2}^{h/2} [Q] \begin{pmatrix} \varepsilon_x \\ \varepsilon_y \\ \gamma_{xy} \end{pmatrix} z\, dz
$$

$$
= \int_{-h/2}^{h/2} [Q] z^2 dz \begin{pmatrix} -w_{,xx} \\ -w_{,yy} \\ -2w_{,xy} \end{pmatrix}
$$

Denoting $\int_{-h/2}^{h/2} [Q] z^2 dz$ as $[D]$, called the flexural rigidity matrix, one has

$$
\begin{pmatrix} M_x \\ M_y \\ M_{xy} \end{pmatrix} = \begin{bmatrix} D_{11} & D_{12} & D_{16} \\ & D_{22} & D_{26} \\ \text{sym.} & & D_{66} \end{bmatrix} \begin{pmatrix} -w_{,xx} \\ -w_{,yy} \\ -2w_{,xy} \end{pmatrix} \tag{9.14}
$$

where one can see the effect of anisotropy as a coupling between the bending and twisting modes (compare with Eq. (2.13)). This coupling is absent for the specially orthotropic configuration.

The equilibrium equations in terms of the stress resultants are the same as for the isotropic plate. The final governing equation is obtained by substitution for the moments from Eq. (9.14) in Eq. (2.17) as

$$
D_{11}w_{,xxxx} + 4D_{16}w_{,xxxy} + 2(D_{12} + 2D_{66})w_{,xxyy}
$$
$$
+ 4D_{26}w_{,xyyy} + D_{22}w_{,yyyy} = q \tag{9.15}
$$

The classical boundary conditions, in terms of w, $w_{,x}$, $w_{,y}$, M_x, M_y, V_x and V_y, remain the same as for the isotropic plate (Sect. 2.5); the natural boundary conditions can be expressed in terms of w as done earlier using the moment–curvature relations of Eq. (9.14). It has to be noted that this results in more complicated expressions than earlier because of bending–twisting coupling; for example, the conditions for a simply supported edge at $x = $ constant are

$$w = M_x = 0$$

i.e. $\quad w = -D_{11}w_{,xx} - D_{12}w_{,yy} - 2D_{16}w_{,xy} = 0$

i.e. $\quad w = -D_{11}w_{,xx} - 2D_{16}w_{,xy} = 0 \qquad (9.16)$

where it should be noted that the enforcement of zero deflection all along the edge automatically implies that the bending curvature $w_{,yy}$ is zero, but has no similar effect on the twist curvature.

Simple Navier or Levy-type solutions are not possible for anisotropic plates; the specially orthotropic plate, however, can be analysed by these methods as is easily seen by comparing the terms in Eq. (9.15) with those of the biharmonic equation (Eq. (2.18)). For example, proceeding exactly as before, one can obtain the Navier solution for the simply supported specially orthotropic plate as

$$w = \sum_m \sum_n \frac{q_{mn} \sin \frac{m\pi x}{a} \sin \frac{n\pi y}{b}}{\pi^4 \left[D_{11}\left(\frac{m}{a}\right)^4 + 2(D_{12} + 2D_{66})\left(\frac{mn}{ab}\right)^2 + D_{22}\left(\frac{n}{b}\right)^4 \right]} \qquad (9.17)$$

in place of the simpler Eq. (4.11).

Using this result, an example is presented below to illustrate the benefits of correct, preferential reinforcement.

Example 9.1 Effect of the Direction of Reinforcement

Consider a rectangular plate with $a = 2b$ subjected to sinusoidal transverse load

$$q = q_o \sin \frac{\pi x}{a} \sin \frac{\pi y}{b}$$

If one decides to use a unidirectionally reinforced orthotropic material for this case with the fibres along the x-direction (called the $(0°)$ plate, the fibre angle being measured anticlockwise with respect to the x-direction), the central deflection of the plate is given by

$$w_{max} = \frac{q_o}{\pi^4 \left[\frac{D_{11}}{(2b)^4} + 2\frac{(D_{12}+2D_{66})}{(2b)^2 b^2} + \frac{D_{22}}{b^4} \right]}$$

If the material properties are $D_{11} = 5D_o, D_{12} + 2D_{66} = D_{22} = D_o$, then the above result reduces to

$$w_{max} = \frac{16q_o b^4}{29\pi^4 D_o} \approx 0.55 \frac{q_o b^4}{\pi^4 D_o}$$

If the same material is used with the fibres oriented along the y-direction (i.e. the $(90°)$ plate), then the values of D_{11} and D_{22} get interchanged, and one ends up with the central deflection as

$$w_{max} = \frac{16q_o b^4}{89\pi^4 D_o} \approx 0.18 \frac{q_o b^4}{\pi^4 D_o}$$

Thus, between the two orientations, the latter is definitely superior. The reason is quite simple. If one uses an isotropic material, it is easy to visualize the curvatures in the x and y directions at the centre of the plate and to recognize that the curvature in the shorter y-direction is larger; thus, it is preferable to use an orthotropic material with greater stiffness in this direction.

9.2 Classical Laminated Plate Theory

Let us consider a composite plate made up of N perfectly bonded layers of different materials or the same fibre-reinforced composite material with different orientations. As a first step, it is necessary to understand how the behaviour of this plate differs qualitatively from that of a homogeneous plate. It is easy to visualize that an arbitrary layered plate would bend under transverse loading in such a way that the top and bottom fibres do not change in length equally; this is because the top half and the bottom half are not, in general, equally stiff. As a consequence, the mid-plane no longer remains unstrained. Thus, if one considers the mid-plane as the reference plane for the development of the plate theory, one has to account for *bending–stretching coupling*, i.e. one has to model mathematically both stretching and bending together even when the plate is subjected to transverse loads alone.

The displacement field of Eq. (2.4) gets modified to

$$
\begin{aligned}
u(x, y, z) &= u_o(x, y) - z w_{,x} \\
v(x, y, z) &= v_o(x, y) - z w_{,y} \\
w(x, y, z) &= w(x, y)
\end{aligned}
\tag{9.18}
$$

where u_o and v_o correspond to stretching of the mid-plane.

The strains at a point (x, y, z) are given by

$$
\begin{aligned}
\varepsilon_x &= u_{o,x} - z w_{,xx} \\
\varepsilon_y &= v_{o,y} - z w_{,yy} \\
\gamma_{xy} &= u_{o,y} + v_{o,x} - 2 z w_{,xy}
\end{aligned}
\tag{9.19}
$$

Here $u_{o,x}$, $v_{o,y}$ and $(u_{o,y} + v_{o,x})$ are the mid-plane normal and shear strains, respectively, denoted by ε_x^o, ε_y^o and γ_{xy}^o; the two bending curvatures and the twist curvature are denoted by κ_x, κ_y and κ_{xy}, respectively. The thicknesswise variation of any of the three total strains of Eq. (9.19) is linear as illustrated in Fig. 9.5.

For convenience, the total strains are written in shortened form as

$$
\{\varepsilon\}_{x-y} = \{\varepsilon^o\} + z\{\kappa\}
\tag{9.20}
$$

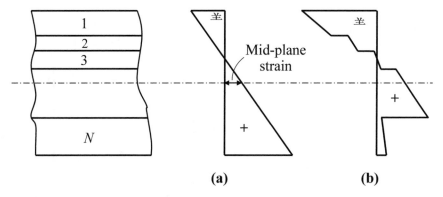

Fig. 9.5 Variation of **(a)** in-plane strains and **(b)** in-plane stresses

where $\{\varepsilon^o\}$ and $\{\kappa\}$ are the appropriately defined mid-plane strain and curvature vectors.

Coming to the stress–strain relationship, each layer is assumed to obey a plane-stress reduced constitutive law; for the k th layer, it is given in the most general form by

$$\{\sigma\}^{(k)}_{x-y} = \left[Q\right]^{(k)}\{\varepsilon\}_{x-y} \qquad (9.21)$$

valid for all possibilities—isotropic, orthotropic or anisotropic– with appropriately defined plane-stress reduced stiffness coefficients. Thus, while the three strains are linear through the entire thickness of the laminate, the stresses are piecewise linear with sudden jumps at the interfaces between the layers (Fig. 9.5).

Regarding stress resultants, one has to introduce in-plane normal and shear forces here, besides bending and twisting moments, because the resulting action of the unsymmetric distribution of the stresses through the thickness is to cause stretching and shearing of the mid-plane along with bending and twisting.

These stress resultants are given by

$$(N_x, N_y, N_{xy}) = \int_{-h/2}^{h/2} (\sigma_x, \sigma_y, \tau_{xy}) dz = \sum_{k=1}^{N} \int (\sigma_x, \sigma_y, \tau_{xy})^{(k)} dz$$

$$(M_x, M_y, M_{xy}) = \int_{-h/2}^{h/2} (\sigma_x, \sigma_y, \tau_{xy}) z dz = \sum_{k=1}^{N} \int (\sigma_x, \sigma_y, \tau_{xy})^{(k)} z dz \qquad (9.22)$$

where the integration over the entire thickness of the plate is carried out in a piecewise manner over the total N layers. N_x and N_y are normal forces per unit length, and N_{xy} is the shear force per unit length. The sign convention for these forces is shown in Fig. 9.6.

Fig. 9.6 Sign
convention—positive forces

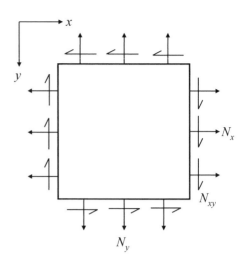

Use of Eqs. (9.20) and (9.21) reduces Eq. (9.22) to the form

$$\left\{ \begin{array}{c} \{N\} \\ \{M\} \end{array} \right\} = \left[\begin{array}{cc} [A] & [B] \\ [B] & [D] \end{array} \right] \left\{ \begin{array}{c} \{\varepsilon^o\} \\ \{\kappa\} \end{array} \right\} \tag{9.23}$$

where $\{N\}$ and $\{M\}$ are the force and moment vectors given by

$$\{N\} = \left\{ \begin{array}{c} N_x \\ N_y \\ N_{xy} \end{array} \right\}, \quad \{M\} = \left\{ \begin{array}{c} M_x \\ M_y \\ M_{xy} \end{array} \right\}$$

and $[A]$, $[B]$ and $[D]$ are symmetric 3×3 matrices defined as

$$[A], [B], [D] = \sum_{k=1}^{N} \int [\overline{Q}]^{(k)} (1, z, z^2) dz \tag{9.24}$$

Equation (9.23) is the constitutive law for the layered plate, i.e. the relation between the generalized forces (in-plane forces and moments) and the generalized displacements (mid-plane strains and curvatures). Matrix $[A]$ relates the in-plane forces to the mid-plane strains and is referred to as the *extensional stiffness* matrix. Matrix $[D]$, relating the moments to the curvatures, is called the *bending stiffness* matrix. Matrix $[B]$, which relates the in-plane forces to the curvatures, and the moments to the mid-plane strains, is called the *bending-stretching coupling stiffness* matrix. For an arbitrary laminate with anisotropic layers, all these three matrices are fully populated; in that case, one can see that there is coupling between all the modes – stretching, shearing, bending and twisting, i.e. the laminate would undergo all these deformations simultaneously due to any general in-plane or transverse load.

Fig. 9.7 Additional forces
on a plate element

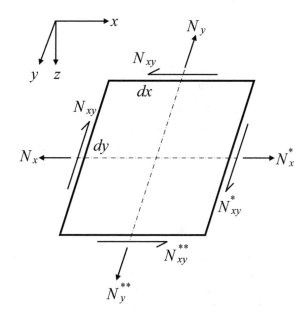

Let us now consider the equilibrium of the layered plate. Besides the moments and the transverse shear forces shown in Fig. 2.3, additional stress resultants N_x, N_y and N_{xy} (and their incremented counterparts) now appear (Fig. 9.7). Since none of these forces acts in the z-direction or contributes to moments about x and y axes, the corresponding equations derived in Sect. 2.3 (Eqs. (2.14–2.16)) are still valid.

The z-direction moment equilibrium can be seen to be satisfied automatically (after neglecting higher-order terms as was done earlier). Thus, only the equilibrium of forces in x and y directions needs to be studied now.

Summation of forces in the x-direction yields

$$- N_x dy + (N_x + N_{x,x}dx)dy - N_{xy}dx + (N_{xy} + N_{xy,y}dy)dx = 0$$
$$\text{i.e.}\quad N_{x,x} + N_{xy,y} = 0 \tag{9.25}$$

Similarly, for the y-direction, one gets

$$N_{xy,x} + N_{y,y} = 0 \tag{9.26}$$

These two equations should now be satisfied along with Eqs. (2.14–2.16), or their equivalent Eq. (2.17) obtained after elimination of Q_x and Q_y,

$$\text{i.e.}\quad M_{x,xx} + 2M_{xy,xy} + M_{y,yy} + q = 0 \tag{9.27}$$

The governing equations are obtained by expressing the above three equations in terms of the displacements using Eqs. (9.19) and (9.23).

They are given by

$$
\begin{aligned}
&A_{11}u_{o,xx} + 2A_{16}u_{o,xy} + A_{66}u_{o,yy} \\
&\quad + A_{16}v_{o,xx} + (A_{12} + A_{66})v_{o,xy} + A_{26}v_{o,yy} \\
&\quad - B_{11}w_{,xxx} - 3B_{16}w_{,xxy} - (B_{12} + 2B_{66})w_{,xyy} - B_{26}w_{,yyy} = 0 \\
&A_{16}u_{o,xx} + (A_{12} + A_{66})u_{o,xy} + A_{26}u_{o,yy} \\
&\quad + A_{66}v_{o,xx} + 2A_{26}v_{o,xy} + A_{22}v_{o,yy} \\
&\quad - B_{16}w_{,xxx} - (B_{12} + 2B_{66})w_{,xxy} - 3B_{26}w_{,xyy} - B_{22}w_{,yyy} = 0 \\
&B_{11}u_{o,xxx} + 3B_{16}u_{o,xxy} + (B_{12} + 2B_{66})u_{o,xyy} + B_{26}u_{o,yyy} \\
&\quad + B_{16}v_{o,xxx} + (B_{12} + 2B_{66})v_{o,xxy} + 3B_{26}v_{o,xyy} + B_{22}v_{o,yyy} \\
&\quad - D_{11}w_{,xxxx} - 4D_{16}w_{,xxxy} - 2(D_{12} + 2D_{66})w_{,xxyy} \\
&\quad - 4D_{26}w_{,xyyy} - D_{22}w_{,yyyy} + q = 0
\end{aligned}
\tag{9.28}
$$

where, one should note, once again, the coupling between u_o, v_o and w.

These equations should be supplemented by boundary conditions. Corresponding to the two additional governing equations here in comparison to the case of homogeneous plates, additional boundary conditions are required in terms of the mid-plane displacements or their force-counterparts. A general statement of the boundary conditions, accounting for all possible classical combinations, is to say that one parameter from each of the following bracketed groups should be specified as zero:

$$
(u_0, N_x); \ (v_0, N_{xy}); \ (w, V_x); \ (w_{,x}, M_x) \text{ at the edge } x = \text{constant}
$$
$$
(u_0, N_{xy}); \ (v_0, N_y); \ (w, V_y); \ (w_{,y}, M_y) \text{ at the edge } y = \text{constant}
\tag{9.29}
$$

It can be noted that four different combinations of in-plane conditions are possible for each of the transverse boundary conditions. For example, corresponding to a simple support at $x = $ constant, one can have

$$
u_0 = v_0 = 0 \ \text{ or } \ u_0 = N_{xy} = 0 \ \text{ or } \ N_x = v_0 = 0 \ \text{ or } \ N_x = N_{xy} = 0 \tag{9.30a}
$$

along with

$$
w = M_x = 0 \tag{9.30b}
$$

The combination to be used should be chosen such that the actual physical restraint is simulated most realistically.

Now that the modified equations of CPT as applicable to a general layered plate have been presented, it is appropriate to consider the special case of a symmetric laminate (Fig. 9.8). A symmetric laminate is one for which the lay-up (specified by the material and the thickness of the individual layers in sequence) of one half of the plate is a mirror image of that of the other half about the mid-plane. For such laminates, [B] becomes a null matrix because all its elements turn out to be integrals

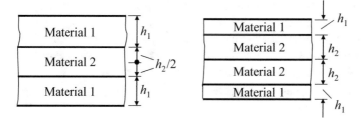

Fig. 9.8 Symmetric laminates

of odd functions of z between the limits $-h/2$ and $h/2$ (see Eq. (9.24)); hence, bending–stretching coupling is absent. Thus, the governing equations get split into two parts - the first two equations in terms of u_o and v_o alone, corresponding to the *stretching problem* (applicable when only in-plane loads are present), and the third equation in terms of w alone corresponding to the *bending problem* (applicable when only transverse loads or moments are present). This bending equation for the symmetric laminate is the same as that of a single anisotropic plate (Eq. (9.15), the only difference being that the in-plane stresses vary in a piecewise manner through the thickness for the former.

Equations (9.25–9.28) pertain to the problem of static flexure under transverse loading. They can be modified by the inclusion of appropriate terms, as was done for isotropic homogeneous plates, for the study of free vibrations and stability. For the dynamic problem, the inertia terms that would come into Eqs. (9.25–9.27), respectively, are $-\bar{\rho}u_{o,tt}$, $-\bar{\rho}v_{o,tt}$ and $-\bar{\rho}w_{,tt}$; of these, the first two are not significant for thin plates and are usually neglected.

For the buckling problem, if the laminate were to be symmetric, the analysis is exactly on the same lines as that of Sect. 7.2; for a rectangular plate subjected to uniform edge loads $\overline{N}_x, \overline{N}_y, \overline{N}_{xy}$ (the overbars being used to distinguish these from the internal force resultants defined in Eq. (9.22)), the governing equation is

$$D_{11}w_{,xxxx} + 4D_{16}w_{,xxxy} + 2(D_{12} + 2D_{66})w_{,xxyy}$$
$$+ 4D_{26}w_{,xyyy} + D_{22}w_{,yyyy} = \overline{N}_x w_{,xx} + \overline{N}_y w_{,yy} + 2\overline{N}_{xy}w_{,xy} \qquad (9.31)$$

If the laminate is unsymmetric, it should first be noted that the problem is not one of bifurcation because any in-plane loading would immediately cause bending due to bending–stretching coupling; this is similar to the problem of the eccentrically loaded column. In such a case, one has to define the critical load as the limiting load at which transverse deflections rapidly grow and tend to infinitely large values. This critical load can be correctly captured by ignoring the prebuckling displacements (but not the bending-stretching coupling itself) and proceeding as for the symmetric laminate, i.e. by considering a small perturbation from the prebuckled flat shape, during which mid-plane displacements u_o and v_o also occur besides w because of coupling, and by studying the equilibrium of this adjacent configuration. The resulting equations are the same as Eq. (9.28) with the last equation having $\overline{N}_x w_{,xx} + \overline{N}_y w_{,yy} + 2\overline{N}_{xy} w_{,xy}$ in place of q for the case of uniform edge loading.

Coming to solutions for laminates, exact analytical solutions are not possible for an arbitrary laminate with anisotropic layers. However, for special cases of rectangular plates, Navier-type solutions are possible. These cases are:

(a) Cross-ply plates, or plates with isotropic layers, with the simple support conditions given by

$$
\begin{aligned}
w = M_x = N_x = v_0 = 0 \text{ at } x = 0, a \\
w = M_y = N_y = u_0 = 0 \text{ at } y = 0, b
\end{aligned}
\tag{9.32}
$$

which can be identified to be of the shear-diaphragm type. A cross-ply plate is one for which all the layers are specially orthotropic (fibre angle $0°$ or $90°$).

(b) Antisymmetric angle-ply plates with the simple support conditions given by

$$
\begin{aligned}
w = M_x = N_{xy} = u_0 = 0 \text{ at } x = 0, a \\
w = M_y = N_{xy} = v_0 = 0 \text{ at } y = 0, b
\end{aligned}
\tag{9.33}
$$

An antisymmetric angle-ply plate is one for which there is a unidirectionally reinforced $-\theta°$ layer on the other side, as the mirror image about the mid-plane, for every $+\theta°$ layer, e.g. a $(15°/30°/-30°/-15°)$ lay-up with the layers listed starting from the top-most one.

These solutions, for static flexure as well as frequency or stability analysis, are discussed in detail in various textbooks on composite structures.[2] We shall discuss here just the static analysis of the cross-ply plate to illustrate the methodology.

Example 9.2 Flexure of an Unsymmetric Cross-Ply Plate

Let us consider, without loss of generality, just one harmonic of the transverse load as given by

$$
q(x, y) = q_{mn} \sin \frac{m\pi x}{a} \sin \frac{n\pi y}{b}
$$

keeping in mind that any general load can be expanded in a sine series and the corresponding solutions can be superposed.

As stated above, the cross-ply plate is made up of specially orthotropic layers and hence

$$
\overline{Q}_{16} = \overline{Q}_{26} = 0 \text{ for all the layers}
$$

[2]see, for example:

   R. M. Jones, Mechanics of Composite Materials, Taylor and Francis, 1999;

   J. M. Whitney, Structural Analysis of Laminated Anisotropic Plates, Technomic, 1987.

Thus, $A_{16} = A_{26} = B_{16} = B_{26} = D_{16} = D_{26} = 0$

As a consequence, many terms in Eq. (9.28) become zero; more specifically, the first equation has just the $u_{o,xx}$, $u_{o,yy}$, $v_{o,xy}$, $w_{,xxx}$ and $w_{,xyy}$ terms, the second has $u_{o,xy}$, $v_{o,xx}$, $v_{o,yy}$, $w_{,xxy}$ and $w_{,yyy}$ terms, and the third has $u_{o,xxx}$, $u_{o,xyy}$, $v_{o,xxy}$, $v_{o,yyy}$, $w_{,xxxx}$, $w_{,xxyy}$ and $w_{,yyyy}$ terms. A careful look at these terms indicates that the equations get simplified to algebraic equations if one seeks the solutions in the form

$$u_o = U \cos \frac{m\pi x}{a} \sin \frac{n\pi y}{b}$$

$$v_o = V \sin \frac{m\pi x}{a} \cos \frac{n\pi y}{b}$$

$$w_o = W \sin \frac{m\pi x}{a} \sin \frac{n\pi y}{b}$$

One can also verify that these assumed variations in the x and y directions lead to automatic satisfaction of the edge conditions of Eq. (9.32) wherein N_x, N_y, M_x and M_y can be expressed in terms of $u_{o,x}$, $v_{o,y}$, $w_{,xx}$ and $w_{,yy}$ using the laminate constitutive law Eq. (9.23). (At this stage, one should also verify that such automatic satisfaction of the edge conditions is not possible with any of the other simple support variants of Eq. (9.30a)).

Thus, by the above Navier-type methodology, the edge conditions are automatically satisfied and the governing equations reduce to a set of non-homogeneous algebraic equations in terms of U, V, W and q_{mn} which can be readily solved.

The solution for the other class of problems, the antisymmetric angle-ply laminates, is quite similar to the above except that the sine and cosine variations of the in-plane displacements get interchanged. Using such solutions, it can be shown that the effect of bending–stretching coupling (or the B_{ij} terms in the governing equations) is quite significant and neglect of this can lead to severe underestimation of the maximum deflection and severe overestimation of the natural frequency and buckling load. In all cases, the effect of this coupling is to reduce the overall stiffness of the plate (because it represents greater flexibility of the structure in terms of the mid-plane displacements)—and thus to increase deflections and to decrease the natural frequencies and buckling loads. Hence, a neglect of this coupling is always non-conservative and may be dangerously so.

Similar to the above Navier solutions, one can also attempt Levy-type solutions for the above two categories of laminates with two opposite edges simply supported and with arbitrary conditions on the other two edges. However, because of multiple governing equations here as compared to the single equation for a homogeneous plate, such solutions become a little complicated.[3] For such cases, an alternative, less tedious double-series solution is also possible.[4]

Problems involving other boundary conditions or lay-ups cannot be solved exactly and one has to resort to approximate analytical or numerical methods. If one wishes to use Rayleigh-Ritz method, the strain energy density function has to be derived, as was done for the homogeneous isotropic plate in Sect.2.8, but now accounting for the various elastic couplings. This is given by

$$
\begin{aligned}
U_e &= \frac{1}{2} \int (\sigma_x \varepsilon_x + \sigma_y \varepsilon_y + \tau_{xy} \gamma_{xy}) d\text{Vol} \\
&= \frac{1}{2} \int [\sigma_x(\varepsilon_x^o + z\kappa_x) + \sigma_y(\varepsilon_y^o + z\kappa_y) + \tau_{xy}(\gamma_{xy}^o + z\kappa_{xy})] d\text{Vol} \\
&= \frac{1}{2} \int [N_x \varepsilon_x^o + M_x \kappa_x + N_y \varepsilon_y^o + M_y \kappa_y + N_{xy}\gamma_{xy}^o + M_{xy}\kappa_{xy}] d\text{Area}
\end{aligned}
$$

which yields, after substitution from Eq. (9.23),

$$
U_e = \frac{1}{2} \int
\begin{bmatrix}
A_{11}u_{o,x}^2 + 2A_{12}u_{o,x}v_{o,y} + A_{22}v_{o,y}^2 \\
+2(A_{16}u_{o,x} + A_{26}v_{o,y})(u_{o,y} + v_{o,x}) \\
+A_{66}(u_{o,y} + v_{o,x})^2 \\
-2B_{11}u_{o,x}w_{,xx} - 2B_{12}(v_{o,y}w_{,xx} + u_{o,x}w_{,yy}) \\
-2B_{22}v_{o,y}w_{,yy} - 4B_{66}w_{,xy}(u_{o,y} + v_{o,x}) \\
-2B_{16}\{w_{,xx}(u_{o,y} + v_{o,x}) + 2u_{o,x}w_{,xy}\} \\
-2B_{26}\{w_{,yy}(u_{o,y} + v_{o,x}) + 2v_{o,y}w_{,xy}\} \\
+D_{11}w_{,xx}^2 + 2D_{12}w_{,xx}w_{,yy} + D_{22}w_{,yy}^2 \\
+4D_{66}w_{,xy}^2 + 4(D_{16}w_{,xx} + D_{26}w_{,yy})w_{,xy}
\end{bmatrix}
d\text{Area} \qquad (9.34)
$$

While using Rayleigh-Ritz method, one has to keep in mind that the rate of convergence of the results may be influenced by the presence of the coupling. For example, consider the free vibration analysis of a symmetric angle-ply plate to determine its flexural frequencies. The strain energy for this problem involves just the D_{ij} terms and the only coupling that arises is bend–twist coupling represented by D_{16} and D_{26} terms. For simply supported boundary conditions, it has been found[5] that a total of

---

[3] see for example:

Sudhakar Sharma, N. G. R. Iyengar, P. N. Murthy, Buckling of antisymmetric cross- and angle-ply laminated plates, International Jl. of Mechanical Sciences, 22, 1980, 607–620;

A. A. Khdeir, Comparison between shear deformable and Kirchhoff theories for bending, buckling and vibration of antisymmetric angle-ply laminated plates, Composite Structures, 13, 1989, 159–172.

[4] K. Bhaskar, An elegant and rigorous analytical solution for anti-symmetric angle-ply rectangular plates with any combination of simply supported and clamped edges, Jl. of Reinforced Plastics and Composites, 25, 2006, 1679–1689.

[5] A. W. Leissa, Y. Narita, Vibration studies for simply supported symmetrically laminated rectangular plates, Composite Structures, 12, 1989, 113–132.

Fig. 9.9 A functionally
graded plate

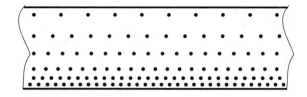

144 terms are required in a double-sine series approximation for the mode shape to
obtain convergent results for the fundamental frequency while a single term would
yield the exact solution if the bend–twist coupling terms were to be absent. Similarly,
with reference to buckling of unsymmetric laminates, it has been reported[6] that a
large number of terms is required in each of the series assumed for u_o, v_o and w to
obtain convergent results.

9.3 CPT for Functionally-Graded Plates

A functionally graded material (FGM) is a composite material made up of two
or more phases with a continuously varying composition – usually a particulate-
reinforced composite with the volume fraction of the particulates varying in one or
more directions, modelled as an isotropic material with continuously varying elastic
and strength properties. For FGM plates, the gradation is usually, but not necessarily,
in the thickness direction (see Fig. 9.9) and this case will be considered here; if the
gradation is in the in-plane directions, the problem is similar to that of a variable
thickness plate (see Chap. 12).

With thickness-direction gradation, the plate behaves just like an unsymmetric
laminate with coupling between bending and stretching with the governing equations
remaining exactly the same. All that is required is the determination of the stiffness
matrices $[A]$, $[B]$ and $[D]$, and this needs a description of the gradation.

For this purpose, the simplest approximation possible is to divide the thickness
into several artificial homogeneous layers and assign an average particulate volume
fraction for each specific layer. The elastic constants E and μ of any such layer are
then obtained by the simple rule of mixtures given by

$$E = E_r V_r + E_m (1 - V_r)$$
$$\mu = \mu_r V_r + \mu_m (1 - V_r) \tag{9.35}$$

where the subscript m refers to the matrix and r to the particulate reinforcement, and
V_r is the volume fraction of the reinforcement.

---

[6]V. J. Papazoglou, N. G. Tsouvalis, G. D. Kyriakopoulos, Buckling of unsymmetric laminates
under linearly varying biaxial in-plane loads combined with shear, Composite Structures, 20, 1992,
155–163.

Alternatively, one can employ a convenient mathematical law with one or more curve-fitting parameters to describe the variation of the volume fraction or the elastic constants themselves with respect to the thickness coordinate. The most commonly used laws are described below.

(a) *Exponential Law*

According to this, the Young's modulus of the FGM is assumed to vary exponentially as given by

$$E(z) = Ae^{B(z+h/2)} \tag{9.36}$$

where the constants A and B can be expressed in terms of E_{top} and E_{bottom}, the values of E at $z = \mp h/2$; these may be obtained using the rule of mixtures if the corresponding volume fractions are known.

In this approach, the Poisson's ratio μ is usually taken to be constant through the entire thickness and equal to some value between μ_m and μ_r which are often nearly the same. This is so that the plane-stress reduced stiffness coefficients, equal to $\frac{E}{(1-\mu^2)}$, etc. (see Eq. (2.9)), also have a simple exponential variation. The shape of this exponential variation is as shown in Fig. 9.10.

A minor and slightly more complicated variant of the above approach is to assume an exponential variation directly for each of the plane-stress reduced stiffness coefficients, given generically as

$$Q_{ij}(z) = Ae^{B(z+h/2)} \tag{9.37}$$

where A and B take different values for different Q_{ij}.

(b) *Power Law*

This is a more general mathematical law as compared to the exponential law in that it has an extra parameter that can be chosen to represent different variations through the thickness, all between the same two values at the extreme points $z = \pm h/2$.

Fig. 9.10 Exponential law variation

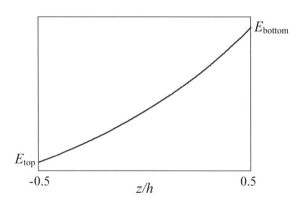

Fig. 9.11 Power law variation

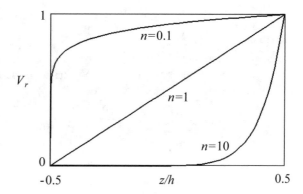

For example, consider a FGM plate with the extreme values of the particulate volume fraction being 1 and 0 (i.e. just the reinforcement phase at the bottom and just the matrix phase at the top). For this plate, the power law variation is expressed as

$$V_r(z) = \left(\frac{z + h/2}{h}\right)^n \tag{9.38}$$

where n is a parameter to be chosen.

If $n = 1$, the variation is linear through the thickness; for much larger or smaller values than 1, the variation shows steep gradients near the top or bottom surface as shown in Fig. 9.11, while for intermediate values one gets smooth non-linear curves falling in between these extremes. Thus, this law can be used to provide a closer approximation to the actual variation as compared to the exponential law. In this approach, once again, the Poisson's ratio is taken as constant through the entire thickness while E is calculated using the rule of mixtures.

As stated earlier, the above laws are the most commonly used though more complicated ones can also be employed.[7] Once these gradation laws are chosen for the material constants, the elements of the [A], [B] and [D] matrices can be calculated by appropriate integration through the thickness; thereafter the problem can be solved as done for unsymmetric laminates.

Though particulate-reinforced FGMs have been commonly considered in the literature, there have also been some studies on fibrous composites with variable fibre spacing,[8] leading to orthotropic behaviour.

---

[7]For an example, see: S. Chi, Y. Chung, Mechanical behaviour of FGM plates under transverse load—Part I: Analysis, International Jl. of Solids and Structures, 43, 2006, 3657–3674.

[8]see, for example:

E. Feldman, J. Aboudi, Buckling analysis of functionally-graded plates subjected to uniaxial loading, Composite Structures, 38, 1997, 29–36;

S. Y. Kuo, L. Shiau, Buckling and vibration of composite laminated plates with variable fiber spacing, Composite Structures, 90, 2009, 196–200.

Summary

The modifications required in the classical isotropic homogeneous thin plate theory to account for in-plane anisotropy or a layered construction or a continuous transverse inhomogeneity is briefly discussed here. In general one or more elastic couplings arise for these cases, and it is important to account for these carefully because any of these has an overall effect of decreasing the stiffness of the plate structure.

Chapter 10
Elasticity Solutions for Plates

As stated in Chap. 1, elasticity solutions for plate problems, wherein the structure is treated as a three-dimensional continuum, are valuable because they provide a standard benchmark against which the accuracy of two-dimensional engineering theories can be judged. Further, they also provide a set of correct displacement and stress variations, the study of which enables the development of approximate theories based on very realistic assumptions.

This has been the main motivation behind the various elasticity solutions available in the literature for isotropic and orthotropic homogeneous plates and laminates. The purpose of this chapter is to give an overview of some such solutions selected from the literature so as to point out the deficiencies of CPT and possible improvements. We shall confine attention to fairly simple problems involving rectangular strips or plates and shall not discuss circular plate problems which are more complicated but lead to essentially the same conclusions about CPT.

10.1 Cylindrical Bending of a Cantilevered Plate Strip Under Tip Shear

We shall start with the plane strain counterpart of the classical elasticity solution for a cantilever subjected to tip shear because the final results for this problem, as compared to all other results available, serve as the most effective statement of the significance of the non-classical shear deformation effect and the parameters on which it depends.

© The Author(s) 2021   159
K. Bhaskar and T. K. Varadan, *Plates*,
https://doi.org/10.1007/978-3-030-69424-1_10

Fig. 10.1 Cantilevered plate
strip

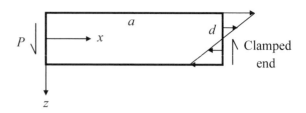

10.1.1   Homogeneous Strip

The plane strain elasticity solution for the problem of the simply supported plate
strip was obtained in Chap. 1 using the displacement approach. For the cantilevered
plate strip under consideration now (Fig. 10.1), we shall use the stress approach.

In the stress approach, one starts by seeking a solution for the stresses by solving
the equilibrium equations along with the boundary conditions and then obtains the
strains and the displacements using the constitutive law and the strain–displacement
relations. While seeking the stress field, the two equilibrium equations (Eq. 1.1a) and
the boundary conditions alone are not adequate in the sense that an infinite number of
solutions would then be possible; a unique solution for the problem can be obtained
only by imposing the additional requirement that the corresponding strain field should
represent a state of deformation wherein the deformed shapes of adjacent elements of
the continuum are compatible, i.e. the elements can be fit together after deformation
as in a jigsaw puzzle. This requirement can be imposed in terms of a single *strain
compatibility equation* for plane stress or plane strain problems as given by[1]

$$\varepsilon_{x,zz} + \varepsilon_{z,xx} = \gamma_{xz,xz} \tag{10.1}$$

for a problem in the x–z plane.

For the present problem, we have

$$\varepsilon_y = 0 = \frac{\sigma_y - \mu(\sigma_x + \sigma_z)}{E}$$

leading to $\sigma_y = \mu(\sigma_x + \sigma_z)$.

---

[1] K. Bhaskar, T. K. Varadan, Theory of Isotropic/Orthotropic Elasticity, Ane Books/CRC Press,
2009.

Hence,

$$\varepsilon_x = \frac{\sigma_x - \mu(\sigma_y + \sigma_z)}{E} = \frac{(1 - \mu^2)}{E}\sigma_x - \frac{\mu(1 + \mu)}{E}\sigma_z \qquad (10.2a)$$

and similarly,

$$\varepsilon_z = -\frac{\mu(1 + \mu)}{E}\sigma_x + \frac{(1 - \mu^2)}{E}\sigma_z \qquad (10.2b)$$

In addition, one has

$$\gamma_{xz} = \frac{\tau_{xz}}{G} = \frac{2(1 + \mu)}{E}\tau_{xz} \qquad (10.2c)$$

Using these, the compatibility equation can be rewritten as

$$[(1 - \mu)\sigma_x - \mu\sigma_z]_{,zz} + [-\mu\sigma_x + (1 - \mu)\sigma_z]_{,xx} = 2\tau_{xz,xz} \qquad (10.3a)$$

which requires to be satisfied along with the following two-dimensional equilibrium equations in the x–z plane:

$$\sigma_{x,x} + \tau_{xz,z} = 0$$
$$\tau_{xz,x} + \sigma_{z,z} = 0 \qquad (10.4)$$

to obtain the stress field.

At this stage, it is possible to simplify Eq. (10.3a) using the equilibrium equations; this is done by differentiating the first of Eq. (10.4) with respect to x and the second with respect to z to obtain, respectively,

$$\tau_{xz,zx} = -\sigma_{x,xx} \text{ and } \tau_{xz,xz} = -\sigma_{z,zz}$$

and hence,

$$2\tau_{xz,xz} = -\sigma_{x,xx} - \sigma_{z,zz}$$

Using this in Eq. (10.3a) and simplifying, one gets

$$\nabla^2(\sigma_x + \sigma_z) = 0 \qquad (10.3b)$$

The solution of the problem governed by Eq. (10.3b) and Eq. (10.4) is usually obtained by using Airy's stress function $\phi(x,z)$ defined as

$$\phi_{,zz} = \sigma_x, \quad \phi_{,xx} = \sigma_z, \quad -\phi_{,xz} = \tau_{xz} \qquad (10.5)$$

so as to satisfy the equilibrium equations directly and to reduce Eq. (10.3b) to the biharmonic equation

$$\nabla^4\phi \equiv \frac{\partial^4\phi}{\partial x^4} + 2\frac{\partial^4\phi}{\partial x^2\partial z^2} + \frac{\partial^4\phi}{\partial z^4} = 0 \qquad (10.6)$$

The solution of a structural problem using this stress function approach is often based on the *inverse method* where different functions for ϕ are examined for their suitability and the correct one identified whenever the problem is amenable to solution by this method. Only a limited number of problems can be solved by this trial and error approach, and the cantilevered strip under tip load is one among them.

For this case (Fig. 10.1), the stress function is

$$\phi = Axz + Bxz^3 \qquad (10.7)$$

where A and B are constants which need to be fixed up such that the following boundary conditions are satisfied.

$$\sigma_z = \tau_{xz} = 0 \text{ at } z = \pm d/2 \text{ for all } x;$$

$$\int_{-d/2}^{d/2} \sigma_x dz = \int_{-d/2}^{d/2} \sigma_x z dz = 0, \quad \int_{-d/2}^{d/2} \tau_{xz} dz = -P \text{ at } x = 0;$$

$$\int_{-d/2}^{d/2} \sigma_x dz = 0, \quad \int_{-d/2}^{d/2} \sigma_x z dz = -Pa, \quad \int_{-d/2}^{d/2} \tau_{xz} dz = -P \text{ at } x = a \qquad (10.8)$$

where P is the tip load per unit length of the strip in the y-direction.

Note that the end conditions at $x = 0$, a are not specified in a point wise sense but only in an integral form; this is because we are interested in any elasticity problem corresponding to the tip-loaded cantilever that is amenable to solution by this inverse method. Further, for now, all the boundary conditions including that of the clamped end are specified in terms of stresses.

With the stresses given by

$$\sigma_x = \phi_{,zz} = 6Bxz$$
$$\sigma_z = \phi_{,xx} = 0$$
$$\tau_{xz} = -\phi_{,xz} = -A - 3Bz^2 \qquad (10.9)$$

one can verify that the solution given by

$$A = \frac{3P}{2d}, \quad B = -\frac{2P}{d^3}$$

satisfies all the boundary conditions and corresponds to the problem where the tip load P is applied as a parabolic distribution of shear stresses given by

$$\tau_{xz} = -\frac{6P}{d^3}\left(\frac{d^2}{4} - z^2\right) \tag{10.10}$$

and the fixed end moment is due to a linear antisymmetric distribution of σ_x about the neutral axis $z = 0$ as shown in Fig. 10.1.

The strain field is given by

$$\varepsilon_x = \frac{(1-\mu^2)}{E}\left(\frac{-12Pxz}{d^3}\right)$$

$$\varepsilon_z = -\frac{\mu(1+\mu)}{E}\left(\frac{-12Pxz}{d^3}\right)$$

$$\gamma_{xz} = -\frac{6P}{Gd^3}\left(\frac{d^2}{4} - z^2\right) \tag{10.11}$$

Integration of ε_x with respect to x yields

$$u = \frac{(1-\mu^2)}{E}\left(\frac{-6Px^2z}{d^3}\right) + f(z)$$

Integration of ε_z with respect to z yields

$$w = -\frac{\mu(1+\mu)}{E}\left(\frac{-6Pxz^2}{d^3}\right) + g(x)$$

Using these expressions to obtain the shear strain, one gets

$$\gamma_{xz} = u_{,z} + w_{,x}$$
$$= \frac{(1-\mu^2)}{E}\left(\frac{-6Px^2}{d^3}\right) + f'(z) - \frac{\mu(1+\mu)}{E}\left(\frac{-6Pz^2}{d^3}\right) + g'(x)$$

If one equates the above shear strain expression with that of Eq. (10.11), one gets

$$\frac{3P}{2Gd} + \frac{(1-\mu^2)}{E}\left(\frac{-6Px^2}{d^3}\right) + g'(x) = \frac{6Pz^2}{Gd^3} - f'(z) + \frac{\mu(1+\mu)}{E}\left(\frac{-6Pz^2}{d^3}\right)$$

which states that a function of x alone is equal to a function of z alone; this is possible only if each of them is equal to a constant, say k_1.

By integration with respect to x or z as appropriate, one gets

$$g(x) = \frac{(1 - \mu^2)}{E}\left(\frac{2Px^3}{d^3}\right) - \frac{3Px}{2Gd} + k_1 x + k_2$$

$$f(z) = \frac{2Pz^3}{Gd^3} + \frac{\mu(1 + \mu)}{E}\left(\frac{-2Pz^3}{d^3}\right) - k_1 z + k_3$$

where the constants k_1 to k_3 have to be obtained from the boundary conditions at the restrained end.

Once again, since this is an inverse solution, these end conditions have to be identified by trial and error[2] and are given by

$$u = w = u_{,z} = 0 \text{ at } x = a, \quad z = 0 \tag{10.12}$$

to yield

$$k_1 = \frac{-6Pa^2(1 - \mu^2)}{Ed^3}, \quad k_2 = \frac{4Pa^3(1 - \mu^2)}{Ed^3} + \frac{3Pa}{2Gd}, \quad k_3 = 0$$

The above solution is exact only if the tip load P is applied as a parabolic distribution of τ_{xz} as in Eq. (10.10); however, by virtue of St. Venant's principle, replacement of this distribution by any other statically equivalent one is expected to produce perturbations of the stress and strain fields only in the near vicinity of the tip and not at distances which are large in comparison with the thickness d. Since the tip deflection is the cumulative result of the deformations of several small elements from the fixed end till the tip, its value is likely to be more or less the same irrespective of how exactly P is applied.

A look at the final equations reveals that u has a cubic variation through the thickness indicating that Kirchhoff–Poisson hypothesis is not really true for this problem; the cubic terms involve G and μ showing that the non-zero transverse shear γ_{xz} and the Poisson effect leading to ε_z are the reasons for the warping of the originally straight normal. Regarding the stress field, one can see that the present solution yields the same results as the thin plate theory—a linear bending stress distribution, a parabolic shear distribution and zero transverse normal stress.

The centreline deflection is given by

$$w = \frac{(1 - \mu^2)}{E}\left(\frac{2Px^3}{d^3}\right) - \frac{6Pa^2 x(1 - \mu^2)}{Ed^3} + \frac{4Pa^3(1 - \mu^2)}{Ed^3} + \frac{3P(a - x)}{2Gd}$$

$$= \left(\frac{Px^3}{6D} - \frac{Pa^2 x}{2D} + \frac{Pa^3}{3D}\right) + \frac{3P(a - x)}{2Gd} \tag{10.13}$$

---

[2] For a more complete discussion, see: K.Bhaskar, T.K.Varadan, Theory of Isotropic/Orthotropic Elasticity, Ane Books/CRC, 2009.

where D is the flexural rigidity.

The first bracketed term is the deflection obtained using CPT, while the second term is an additional deflection that occurs due to inclusion of transverse shear strain γ_{xz}, as seen by the presence of G. If these are referred to as the bending deflection w_b and the shear deflection w_s, respectively, then their ratio at the tip of the cantilever is given by

$$\frac{w_s}{w_b} = \frac{9D}{2da^2G} = \frac{3}{8}\left\{\frac{E/(1-\mu^2)}{G}\right\}\left(\frac{d}{a}\right)^2 \tag{10.14}$$

This is an important result that clearly shows the dependence of the shear deflection on the material and geometric parameters.

The thickness-to-length ratio appears to be the more significant factor here because the shear deflection varies quadratically with respect to it and also because the other bracketed term, which can be rewritten as $\left(\frac{2}{1-\mu}\right)$, is not expected to vary much from one metal to another. However, if one looks at this term as the ratio of the in-plane Young's modulus to the transverse shear modulus and considers the possibility of the isotropic material being replaced by one reinforced by fibres along x, then its importance becomes immediately clear. It is easy to visualize that such reinforcement would cause a multi-fold increase in the x-direction stiffness while leaving the transverse shear stiffness more or less unaffected. For example, for graphite-epoxy fibre composites E_L/G_{LT} is of the order of 25–80, while the counterpart for metals is usually around 2.6; thus, for the same thickness-to-length ratio (d/a), the results by CPT are likely to be much more in error for the former than the latter. (It is actually possible to develop, on exactly the same lines as above, the elasticity solution for an orthotropic cantilever and to show that (E_L/G_{LT}) takes the place of E/G—this is left as an exercise for the reader.)

Finally, it should be kept in mind that the CPT estimate of the deflection is an underestimation, and hence its use leads to an unsafe, non-conservative design. This is true when it is applied for the study of free vibration frequencies and buckling loads also, in which case one ends up with over-predicted estimates.

10.1.2 Laminated Strip

Let us now extend the above solution to the case of a symmetric laminate with three isotropic layers (Fig. 10.2) and draw some important conclusions regarding shear deformation.

Assuming that the stress functions of the different layers would be similar to that of the homogenous plate, we start with

$$\phi_1 = A_1xz + B_1xz^3$$
$$\phi_2 = A_2xz + B_2xz^3 \tag{10.15}$$

Fig. 10.2 Cantilevered
laminate

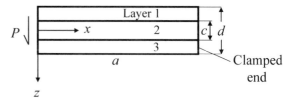

and $\phi_3 = \phi_1$ in view of the expected symmetry/antisymmetry of stresses about the centroidal axis; the symmetry considerations will be used further while discussing the stress and displacement fields and the boundary conditions.

In addition to the boundary conditions of Eq. (10.8), we now have the continuity conditions at the interfaces, which can be stated as

$$\tau_{xz}^{(1)} = \tau_{xz}^{(2)}, \quad \sigma_z^{(1)} = \sigma_z^{(2)} \text{ at } z = -c/2 \text{ for all } x \tag{10.16a}$$

$$u^{(1)} = u^{(2)}, \quad w^{(1)} = w^{(2)} \text{ at } z = -c/2 \text{ for all } x \tag{10.16b}$$

where the former represent equilibrium at the interfaces while the latter, the existence of a perfect bond between the layers.

The stress field is given by

$$\sigma_x^{(1)} = 6B_1 x z$$
$$\sigma_z^{(1)} = 0$$
$$\tau_{xz}^{(1)} = -A_1 - 3B_1 z^2 \tag{10.17}$$

for the first layer and three similar equations for the second layer in terms of A_2 and B_2.

Proceeding further, exactly as was done for the homogeneous strip, one gets three additional constants per layer in the displacement field. If one assumes the same Poisson's ratio for all the layers, this displacement field can be written as

$$u^{(1)} = \frac{3B_1 x^2 z (1 - \mu^2)}{E_1} - \frac{B_1 z^3}{G_1} + \frac{\mu(1 + \mu)}{E_1} B_1 z^3 - \alpha_1 z + \eta_1$$

$$w^{(1)} = -\frac{3B_1 x z^2 \mu(1 + \mu)}{E_1} - \frac{B_1 x^3 (1 - \mu^2)}{E_1} - \frac{A_1 x}{G_1} + \alpha_1 x + \beta_1 \tag{10.18}$$

for the first layer, and similarly for the second layer in terms of A_2, B_2, E_2, G_2 and three new constants α_2, β_2 and η_2.

There are thus a total of ten undetermined constants, and these have to be evaluated using the conditions of Eq. (10.8) and (10.16) along with appropriate clamped end conditions at $x = a$. The continuity of σ_z across the interfaces is automatically satisfied since this stress is identically zero in all the layers. The continuity conditions of the displacements as stated in Eq. (10.16b) actually result in five equations since one has to equate the corresponding coefficients of each power of x on the left-hand side and right-hand side of each of these equations.

Assuming that the clamped end conditions of Eq. (10.12) are suitable for this case also and proceeding further, one can show that all the constants can be evaluated. They are obtained as

$$B_2 = \frac{2P}{c^3(r-1) - d^3r}; \quad B_1 = r B_2; \quad A_1 = -\frac{3B_1 d^2}{4};$$

$$A_2 = A_1 + \frac{3c^2}{4}(B_1 - B_2); \quad \beta_1 = \beta_2 = \frac{A_2 a}{G_2} - \frac{2B_2 a^3 (1 - \mu^2)}{E_2};$$

$$\alpha_2 = \frac{3B_2 a^2 (1 - \mu^2)}{E_2}; \quad \alpha_1 = \alpha_2 + \frac{A_1}{G_1} - \frac{A_2}{G_2};$$

$$\eta_1 = \frac{c(\alpha_1 - \alpha_2)}{2}; \quad \eta_2 = 0 \tag{10.19}$$

where $r = \frac{E_1}{E_2} = \frac{G_1}{G_2}$.

The deflection corresponding to the tip of the centreline is given by

$$w(0, 0) = \frac{4P a^3 (1 - \mu^2)}{E_2[d^3 r - c^3(r-1)]} + \frac{3P a [d^2 r - c^2(r-1)]}{2G_2 [d^3 r - c^3(r-1)]} \tag{10.20}$$

wherein the first term corresponds to the bending component w_b and the second, the shear component w_s. The bending deflection can be expressed in terms of effective flexural rigidity of the laminate given by

$$D_{\text{eff}} = \int \frac{E}{(1 - \mu^2)} z^2 dz = 2\left[\int_0^{c/2} \frac{E_2}{(1 - \mu^2)} z^2 dz + \int_{c/2}^{d/2} \frac{E_1}{(1 - \mu^2)} z^2 dz\right]$$

$$= \frac{E_2}{12(1 - \mu^2)}[d^3 r - c^3(r-1)] \tag{10.21}$$

to yield the familiar result $\frac{P a^3}{3 D_{\text{eff}}}$.

The ratio of the two deflections is given by

$$
\begin{aligned}
\frac{w_s}{w_b} &= \frac{3}{8} \left\{ \frac{E_2/(1-\mu^2)}{G_2} \right\} \frac{[d^2 r - c^2(r-1)]}{a^2} \\
&= \frac{3}{4(1-\mu)} \frac{[d^2 r - c^2(r-1)]}{a^2}
\end{aligned}
\tag{10.22}
$$

which reduces to its homogeneous counterpart (Eq. (10.14)) when r is taken as 1.

At this stage, let us consider some practically important laminate configurations and draw conclusions about the shear deformation effect. For convenience, we shall assume $\mu = 0.25$ in all the cases so that $\frac{3}{4(1-\mu)}$ takes the value of 1; corresponding to this, the ratio of deflections for the homogeneous plate is simply $(d/a)^2$.

Case (a): c = d/3
This case is representative of laminates with more or less equal thickness layers. The conclusions drawn from this case are also reasonably valid for cross-ply laminates with orthotropic composite layers; if the lay-up is taken as $(0°/90°/0°)$, r would correspond to E_L/E_T.

For this case, we have

$$
\begin{aligned}
\frac{w_s}{w_b} &= \frac{[d^2 r - d^2(r-1)/9]}{a^2} \\
&= \frac{(8r+1)}{9} \left(\frac{d}{a} \right)^2 \\
&\approx 0.9r \left(\frac{d}{a} \right)^2 \quad \text{for } r > 2
\end{aligned}
$$

Thus, one can see that, for $r > 2$, the ratio of deflections is more than that for the homogeneous plate by a factor nearly equal to r.

Case (b): c/d = 0.9
This is representative of sandwich plates where two thin stiff facing sheets are separated by a thick weak core; the ratio r in this case can be quite large, ranging from about 100 to 1000 or even more.

For this case,

$$
\frac{w_s}{w_b} = \frac{[d^2 r - c^2(r-1)]}{a^2} \approx \frac{r(d^2 - c^2)}{a^2} = 0.19r \left(\frac{d}{a} \right)^2
$$

which can be many times larger than its counterpart for the homogeneous plate.

Thus, for layered configurations with stiffer outer layers, the shear deflection is of much greater significance than for the homogeneous plate of the same thickness-to-span ratio. In all such cases, it is very important to distinguish between a 'geometrically thin' and a 'technically thin' plate and to employ CPT only for the latter.

Fig. 10.3 Zigzag shape of the deformed normal

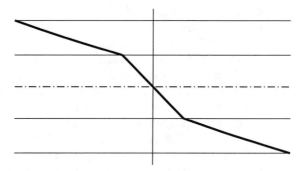

Another interesting result from the above solution is the variation of the in-plane displacement through the thickness, for this provides the shape of the deformed normal. While this variation is nearly linear through the entire thickness for fairly low values of r and d/a indicating that Kirchhoff–Poisson hypothesis is true for such plates, the departure from linearity is significant when r and d/a are increased; the typical variation is zigzag with sharp changes of slope at the interfaces as shown in Fig. 10.3 for the three-layered plate, with a nearly linear variation within each layer.

10.2 Flexure of Simply Supported Rectangular Plates/Laminates Due to Transverse Loading

The solution for this problem is a rather straightforward extension of that for the infinite strip presented in Chap. 1. Consider the plate to be bounded by the edges at $x = 0, a$ and $y = 0, b$ and the lateral surfaces $z = \pm h/2$. Assuming shear-diaphragm-type supports at all the four edges, one has

$$w = \sigma_x = v = 0 \text{ at } x = 0, a \text{ for all } y \text{ and } z$$

$$w = \sigma_y = u = 0 \text{ at } y = 0, b \quad \text{for all } x \text{ and } z \tag{10.23}$$

These boundary conditions can be satisfied exactly if a solution for the displacements is sought as

$$u = \sum_m \sum_n U_{mn}(z) \cos \frac{m\pi x}{a} \sin \frac{n\pi y}{b}$$

$$v = \sum_m \sum_n V_{mn}(z) \sin \frac{m\pi x}{a} \cos \frac{n\pi y}{b}$$

$$w = \sum_m \sum_n W_{mn}(z) \sin \frac{m\pi x}{a} \sin \frac{n\pi y}{b} \tag{10.24}$$

The above step is similar to that of Eq. (1.8) except that a complete series is employed here to take care of any transverse load expressible in a double sine series.

The three equilibrium equations (Eq. 1.1), when written in terms of displacements, turn out to be such that the first equation has terms involving $u_{,xx}$, $u_{,yy}$, $u_{,zz}$, $v_{,xy}$ and $w_{,xz}$; the second has terms involving $u_{,xy}$, $v_{,xx}$, $v_{,yy}$, $v_{,zz}$ and $w_{,yz}$; and the third has terms involving $u_{,xz}$, $v_{,yz}$, $w_{,xx}$, $w_{,yy}$ and $w_{,zz}$. A careful examination of these terms shows that the solution forms chosen in Eq. (10.24) would lead to cancellation of the trigonometric terms in x and y in each of the three equations, thereby reducing the problem to a set of three coupled linear ordinary differential equations in terms of $U_{mn}(z)$, $V_{mn}(z)$ and $W_{mn}(z)$ for each (m, n) combination. These form a sixth-order system and have a general solution in terms of six undetermined constants for each (m,n) combination.

These constants are evaluated by enforcing the lateral surface boundary conditions given by

$$\tau_{xz} = \tau_{yz} = 0 \text{ at } z = \mp h/2 \text{ (top/bottom surfaces) for all } x, y$$
$$\sigma_z = -q_{\text{top}}(x, y) \text{ at } z = -h/2$$
$$\sigma_z = q_{\text{bottom}}(x, y) \text{ at } z = h/2 \tag{10.25}$$

where q_{top} and q_{bottom} are transverse loads per unit area, taken positive when acting in the downward (positive z) direction.

When these lateral surface conditions are expressed in terms of displacements along with the use of Eq. (10.24) and with the applied loads q_{top} and q_{bottom} expanded in double sine series in x and y directions, they reduce to a set of six equations for each (m,n) combination. Thus, the problem becomes a straightforward two-point boundary value problem in the z-direction.

The above methodology is useful for both isotropic and specially orthotropic homogeneous plates. It can also be extended to laminates with isotropic and specially orthotropic layers in which case a general solution is developed for each layer on the above lines in terms of six undetermined constants, and all the constants are evaluated using the top and bottom surface conditions and the conditions corresponding to continuity of the displacements and the transverse stresses (τ_{xz}, τ_{yz} and σ_z) at the interfaces. The only difference in the application of the methodology to isotropic and orthotropic plates arises while deriving the general solution as a combination of six linearly independent solutions; the nature of these component solutions depends on whether the roots of the associated auxiliary equation are real or complex, and whether they are distinct or repeated, and this depends on the material properties. We shall not go into such mathematical details here, for our interest lies more in the physical behaviour that is revealed by applying the three-dimensional elasticity approach, and a proper assessment of the thin plate theory against this standard benchmark. In accordance with this viewpoint, we shall merely present a brief discussion of such salient features, citing appropriate literature that can be referred to for complete solutions as well as extensive numerical results.

Isotropic homogeneous and laminated plates were studied by Srinivas et al.[3] Homogeneous rectangular plates of different aspect ratios and square three-layered symmetric laminates with different relative stiffnesses of the layers were considered. The transverse load was taken to be uniformly distributed on the top surface. It was found that the errors of CPT were larger for deflections than for stresses. While these errors were found to be quite small for homogeneous plates with thickness ratio (thickness/smaller side of the rectangular plan-form) less than 0.1, they were found to be strongly influenced by the relative stiffness ratio ($E_{\text{outer}}/E_{\text{inner}}$) for laminates and to be more severe for plates with stiffer outer layers. From the numerical results, it was concluded that the error in the maximum deflection was roughly proportional to this stiffness ratio, which is consistent with the observations of Sect. 10.1.2 with reference to the cantilevered laminated strip.

Elasticity solutions for specially orthotropic plates and laminates were developed by Srinivas and Rao,[4] and Pagano[5]; while Srinivas and Rao considered homogeneous and sandwich-type plates made up of aragonite crystals with a low degree of orthotropy ($E_x/E_y \approx 2$, $E_x/G_{xz} \approx 6$), Pagano obtained his results for a much higher degree of orthotropy ($E_x/E_y = 25$, $E_x/G_{xz} = 50$), typical of unidirectionally reinforced graphite-epoxy composites, considering a sandwich construction as well as cross-ply lay-ups. These investigations also show that the thin plate theory estimates of the stresses are in general more accurate than the deflection estimates. This is true not just of the in-plane stresses but also the transverse stresses τ_{xz}, τ_{yz} and σ_z obtained as statically equivalent estimates by integration of the three-dimensional equilibrium equations as explained in Sect. 2.6; note that this integration has to be done layer by layer for a laminate with the continuity of interlaminar stresses imposed at every interface, and that the estimation of the transverse stresses is very important for laminates because high values of these stresses lead to delamination failures in a layered construction.

Regarding the displacement variations through the thickness, a zigzag variation was reported for the in-plane displacements with sharp changes of slope at every interface, while the transverse displacement variation was found to be insignificant even for thick plates. Finally, the influence of a high stiffness mismatch between layers was emphasized once again, and the larger errors of the thin plate theory for the sandwich construction, than for the cross-ply lay-up, were attributed to this factor.

While the above-mentioned investigations were for uniform or sinusoidally distributed loads, the more severe cases of localized strip and patch loads were considered, respectively, by Pagano and Wang[6] with reference to cylindrical bending

[3]S. Srinivas, A. K. Rao, C. V. Joga Rao, Flexure of simply supported thick homogeneous and laminated rectangular plates, ZAMM – Jl. of Applied Mathematics and Mechanics, 49(8), 1969, 449–458.

[4]S. Srinivas, A. K. Rao, Bending, vibration and buckling of simply supported thick orthotropic rectangular plates and laminates, International Jl. of Solids and Structures, 6, 1970, 1463–1481.

[5]N. J. Pagano, Exact solutions for rectangular bidirectional composites and sandwich plates, Jl of Composite Materials, 4, 1970, 20–34.

[6]N. J. Pagano, A. S. D. Wang, Further study of composite laminates under cylindrical bending, Jl. of Composite Materials, 5, 1971, 521–528.

of a cross-ply plate, and Savithri and Varadan[7] with reference to a finite rectangular cross-ply plate. These investigations reveal that the thin plate theory errors are significantly more for localized loads than for smoothly varying loads, and further that the deflection w varies significantly through the thickness in the neighbourhood of the localized load.

A further investigation of cross-ply plates subjected to localized moment loads,[8] typical of those arising from attachments, reveals that the classical plate theory estimates for the deflections and stresses in the vicinity of the load are acceptable only for very thin plates with a thickness ratio of 1/500 or less.

It is appropriate to point out at this stage that the methodology employed for the above problems is also useful to study cylindrical bending of plates with off-axis layers (i.e. $\theta \neq 0°$ or $90°$), characterized by a monoclinic constitutive law, and with simply supported edges that freely permit the in-plane tangential displacement. Analysis of such plates[9] leads to essentially the same conclusions as those listed above for orthotropic plates.

A common observation from the above studies is that the thickness-wise variation of the transverse deflection is very small except in the close vicinity of localized loading, and this is true even for plates which have to be categorized as very thick on the basis of severely under-predicted CPT estimates of the mid-plane value. Thus, as the plate becomes thicker, the influence of transverse shear flexibility appears to be more significant than that of stretching or contraction of the normals. The relative importance of these two non-classical effects, conveniently referred to as *shear deformation* and *thickness stretch*, respectively, can be correctly gauged by comparing the corresponding contributions to the total strain energy. Noting that these contributions were obtained as U_s and U_t (see Eq. (1.14)) with reference to the cylindrical bending problem of Sect. 1.3, a plot of their variation with the thickness-to-span ratio would provide the required insight. This is shown in Fig. 10.4 in a convenient normalized form and with μ taken as 0.3.

From this figure, one can clearly see that the shear deformation effect is much more significant than the thickness-stretch effect, and thus one may choose to account for the former alone and continue to neglect the latter while considering improvements to the classical theory.

---

[7]S. Savithri, T. K. Varadan, Laminated plates under uniformly distributed and concentrated loads, ASME Jl. of Applied Mechanics, 59, 1992, 211–214.

[8]T. K. Varadan, K. Bhaskar, Elasticity solutions for cross-ply plates under localized surface moments, Jl. of the Aeronautical Society of India, 53(3), 2001, 154–159.

[9]N. J. Pagano, Influence of shear coupling in cylindrical bending of anisotropic laminates, Jl. of Composite Materials, 4, 1970, 330–343.

Fig. 10.4 Energy plots for the infinite strip of Sect. 1.3

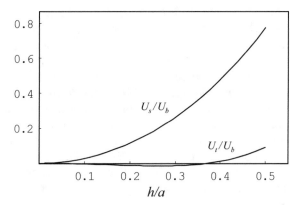

10.3 Vibrations and Stability of Simply Supported Rectangular Plates and Laminates

Using the three-dimensional equations of elastodynamics, one can analyse a rectangular isotropic/specially orthotropic homogeneous or layered plate with shear-diaphragm-type edges; the procedure is essentially the same as that of the static flexure problem in that harmonic variations are assumed for the displacements in the in-plane directions such that the edge conditions are satisfied a priori and the thickness-direction variation is obtained from the governing equations.

Enforcement of the lateral surface and interface conditions yields a non-linear eigenvalue problem of the form

$$[A(\Omega)]\{x\} = \{0\} \tag{10.26}$$

where each element of the matrix $[A]$ depends on the unknown frequency parameter Ω in terms of complicated expressions involving exponential functions. This eigenvalue problem can be solved iteratively using Newton–Raphson or other approximate methods for any given numerical data of material properties and plate dimensions.

Such a solution for isotropic homogeneous plates and symmetric laminates made up of isotropic layers was obtained by Srinivas et al.[10] It was shown that the flexural frequencies predicted by CPT are always overestimates and that the errors are more for higher frequencies characterized by mode shapes with many half-waves. For homogeneous plates, CPT predictions were shown to be quite accurate for modes with $\sqrt{\left(\frac{m\pi h}{a}\right)^2 + \left(\frac{n\pi h}{b}\right)^2} \leq \frac{1}{5}$ where m and n are the number of half-waves in x and y directions, respectively; for laminated plates, the errors are higher in general, especially when the outer plies are stiffer.

---

[10]S. Srinivas, C. V. Joga Rao, A. K. Rao, An exact analysis for vibration of simply-supported homogeneous and laminated thick rectangular plates, Jl. of Sound and Vibration, 12(2), 1970, 187–199.

For cross-ply laminates, frequencies corresponding to cylindrical bending of infinitely long strips were studied by Jones[11]; attention was confined to unsymmetric two-layer plates alone because of the analytical complexity in the case of more layers. The normal modes of vibration of an unsymmetric laminate can be classified into two categories—predominantly flexural and predominantly extensional—even though all of them are coupled modes. It was found by Jones that the flexural frequency predictions of CPT were in general more accurate than the extensional frequency predictions and the errors increased with the number of half-waves, the thickness ratio of the plate and the degree of orthotropy as expected. The errors were found to decrease by the inclusion of in-plane inertia and rotary inertia (the $u_{o,tt}$ and $w_{,xtt}$ terms) in the CPT equations, but only marginally. The modal displacements and stresses of CPT, especially for the extensional modes, were found to be very poor approximations.

Noor[12] studied vibrations of multi-layered simply supported square cross-ply plates; as pointed out above, the exact dynamic analysis of laminates with many layers becomes quite tedious, and hence a numerical solution was obtained in this case using a mixed finite-difference scheme. This study showed that the errors of CPT depend not only on the thickness ratio and the degree of orthotropy, but also on the number of layers and their stacking sequence. The main source of this error was identified to be the shear deformation effect and not rotary inertia which was found to be of negligible importance.

Noor and Burton[13] examined the influence of off-axis layers by considering the specific problem of an antisymmetrically laminated plate; the mathematical formulation is more complicated than that for the orthotropic plate and involves a linear combination of two sets of harmonic variations in the x and y directions. An important conclusion from this study, apart from the conclusions mentioned above with reference to other investigations, is regarding the relative significance of the strain energy components U_b due to bending, U_s due to transverse shear and U_t due to thickness stretch defined as done in Sect. 1.3 [see Eq. (1.13)], but with extra terms in U_b and U_s now to account for bidirectional bending. Considering a fairly high degree of orthotropy (E_L/E_T), it was shown that U_t is very small even for very thick plates while U_s cannot be neglected unless the thickness ratio is quite small; thus, in general, the total strain energy can be considered as the sum of U_b and U_s alone with the latter becoming more significant as the thickness ratio increases.

The buckling problem for isotropic/specially orthotropic plates can also be solved using the theory of three-dimensional elasticity for the case of uniform uniaxial or biaxial compression, the problem being formally identical to that of free vibrations with just a redefinition of the eigenvalue parameter.[14] For laminates, the edge stresses

[11]A. T. Jones, Exact natural frequencies for cross-ply laminates, Jl. of Composite Materials, 4, 1970, 476–491.

[12]A. K. Noor, Free vibrations of multilayered composite plates, AIAA Jl., 11, 1973, 1038–1039.

[13]A. K. Noor, W. S. Burton, Three-dimensional solutions for antisymmetrically laminated anisotropic plates, ASME Jl. of Applied Mechanics, 57, 1990, 182–188.

[14]S. Srinivas, A. K. Rao, Buckling of thick rectangular plates, AIAA Jl., 7, 1969, 1645–1646.

are taken as either uniform across all the layers[15] or distributed between the layers such that ε_x and ε_y are uniform throughout.[16] These solutions show that the buckling load is always over-predicted by CPT and the errors depend on various parameters as described above with reference to vibration frequencies; in certain cases, even the buckled shape is not correctly captured by CPT.

10.4   Solutions for Rectangular Plates with Other Edge Conditions

The analysis of finite rectangular plates with boundary conditions other than shear-diaphragm-type simple supports is not easy because the problem cannot be reduced to a system of ordinary differential equations by assuming suitable expansions satisfying all the edge conditions.

This is clear from the early work of Srinivas and Rao[17] who studied flexure of isotropic homogeneous plates with non-shear-diaphragm-type simple supported edges and clamped edges; each displacement was expanded as a sum of three series, one with harmonic variations in the x, y directions, the second with harmonic variations in the y, z directions and the third with harmonic variations in the x, z directions. While some of the edge conditions were satisfied exactly, others were satisfied only approximately using collocation or orthogonalization. Results were presented for deflections, but not for stresses presumably because of slow convergence; instead, by integrating stresses through the thickness, bending moments were obtained, and these were found to converge quickly. These results show that the thin plate theory errors vary significantly with respect to different boundary conditions; for a square plate with clamped edges the error in the deflection was about four times larger than that corresponding to shear-diaphragm-type edges.

A similar method was employed by Vel and Batra[18] to study orthotropic homogeneous and laminated plates with various boundary conditions including free edges. An important conclusion from this investigation was that a boundary layer characterized by steep stress gradients exists near clamped and free edges while such a phenomenon is absent for shear-diaphragm-type edges. The width of this boundary layer from the edge is typically one order smaller than the total thickness of the plate.

[15]A. Chattopadhyay, H. Gu, Exact elasticity solution for buckling of composite laminates, Composite Structures, 50, 2000, 29–35.

[16]S. Srinivas, A. K. Rao, Bending, vibration and buckling of simply supported thick orthotropic rectangular plates and laminates, International Jl. of Solids and Structures, 6, 1970, 1463–1481.

[17]S. Srinivas, A. K. Rao, Flexure of thick rectangular plates, ASME Jl. of Applied Mechanics, 40, 1973, 298–299.

[18]S. S. Vel, R. C. Batra, Analytical solution for rectangular thick laminated plates subjected to arbitrary boundary conditions, AIAA Jl., 37, 1999, 1464–1473.

Using essentially the same technique, Okumura and Oguma[19] analysed a clamped isotropic homogeneous plate under central patch load and showed that the normal stress distribution corresponding to the restraining moment at any fixed edge is very non-linear through the thickness with infinitely large values at top and bottom. Correspondingly, at the edges, the transverse shear stress variation through the thickness also displays sudden steep gradients near the top and bottom surfaces. These unusual variations are due to complete suppression of all the displacements at the rigidly clamped edges, and they can be captured only by a three-dimensional elasticity solution.

10.5 Corner Reactions in Simply Supported Plates—Insight Obtained From Elasticity Solutions

Within the purview of CPT, consider a simply supported rectangular plate subjected to a simple sinusoidal transverse load given by

$$q = q_o \sin \frac{\pi x}{a} \sin \frac{\pi y}{b} \tag{10.27}$$

The Navier solution for this is given by

$$w = \frac{q_o}{D\pi^4 \left(\frac{1}{a^2} + \frac{1}{b^2}\right)^2} \sin \frac{\pi x}{a} \sin \frac{\pi y}{b} \tag{10.28}$$

with the shear forces at the edges, acting upward, as given by

$$|Q_x| = \frac{q_o}{a\pi \left(\frac{1}{a^2} + \frac{1}{b^2}\right)} \sin \frac{\pi y}{b} \quad \text{along } x = 0, a$$

$$|Q_y| = \frac{q_o}{b\pi \left(\frac{1}{a^2} + \frac{1}{b^2}\right)} \sin \frac{\pi x}{a} \quad \text{along } y = 0, b \tag{10.29}$$

Integrating these forces along the corresponding edges, the total vertical reaction turns out to be $(4q_o ab/\pi^2)$ which is equal to the integral of the load q over the area of the plate (see Problem 3 of Chap. 4). Thus, Q_x and Q_y can be looked upon as the simply supported edge reactions.

Alternatively, since the free edge boundary condition corresponding to a completely unrestrained deflection is in terms of the Kirchhoff's shear force V_x or V_y, one can argue that it is this force that acts to restrain w in the case of simply

---

[19]I. A. Okumura, Y. Oguma, Series solutions for a transversely loaded and completely clamped thick rectangular plate based on the three-dimensional theory of elasticity, Archive of Applied Mechanics, 68, 1998, 103–121.

supported and clamped edges. As part of this argument, one has to include concentrated forces at the four corners; these arise when the twisting moment M_{xy} acting on a small element close to the corner is replaced by two transverse forces acting at its extremities (see Fig. 2.5). Such a contribution comes from both the edges meeting at a corner, and a careful examination reveals that they are in the same direction yielding a total magnitude of $2M_{xy}$ for the corner reaction, usually designated by R.

Thus, the vertical reactions for the simply supported plate under consideration can be taken to include the four distributed Kirchhoff shear forces and the four corner forces. (This also follows from a variational derivation of the classical plate theory where the boundary conditions are defined in terms of a set of generalized displacement–force pairs with one quantity from each pair to be specified; in that case, one ends up with w along the edges paired with V_x or V_y, and w at the corners paired with $2M_{xy}$.)

For the present rectangular plate with load given by Eq. (10.27),

$$M_{xy} = -D(1 - \mu)w,_{xy} = \frac{-q_0(1 - \mu)}{\pi^2 ab\left(\frac{1}{a^2} + \frac{1}{b^2}\right)^2} \cos \frac{\pi x}{a} \cos \frac{\pi y}{b} \qquad (10.30)$$

which is negative at the corners $(0,0)$, (a,b) and positive at the other two corners. With the sign of convention of Fig. 2.2 and replacing the moments by couples of forces as in Fig. 2.5, it is clear that the corner forces will all be downward and of the same magnitude as given by

$$|R| = \frac{-2q_0(1 - \mu)}{\pi^2 ab\left(\frac{1}{a^2} + \frac{1}{b^2}\right)^2} \qquad (10.31)$$

The Kirchhoff shear forces acting upward along the edges can be calculated to be

$$|V_x| = \frac{q_0 ab^2\{a^2(2 - \mu) + b^2\}}{\pi(a^2 + b^2)^2} \sin \frac{\pi y}{b} \quad \text{along } x = 0, a$$

$$|V_y| = \frac{q_0 a^2 b\{a^2 + b^2(2 - \mu)\}}{\pi(a^2 + b^2)^2} \sin \frac{\pi x}{a} \quad \text{along } y = 0, b \qquad (10.32)$$

Integrating these forces along the edges and adding the corner forces algebraically, one gets the total load $(4q_0 ab/\pi^2)$ once again. Thus, the upward V_x and V_y along the edges and the downward R at the corners also balance the total applied load. In support of this argument, the case of a plate freely resting on a rectangular frame can be cited where it is easy to visualize the lifting of the corners when the plate is pushed down at the centre; thus, the corner forces have to be downward to restrain the deflection.

Though the above argument follows from a rigorous variational procedure and is justifiable from the example of the freely resting plate, it is not intuitively acceptable even with reference to this example because a gradually varying distribution of the reaction along each edge from an upward force at the middle to a downward force at the corners appears to be more reasonable than an upward distribution all along the length along with concentrated downward forces at the corners.

There have also been instances[20] wherein the reaction forces, as obtained from a more rigorous theory than CPT, were shown to be quite different from the CPT distributions of V_x and V_y along with R, and closer to those of Q_x and Q_y.

For a satisfactory resolution of the above dilemma—whether the reaction forces of the simply supported plate should be taken as Q_x, Q_y, or the V_x, V_y, R combination, one needs to look at the results of a three-dimensional elasticity solution for the simply supported plate.

Within the purview of such a three-dimensional theory, two simple support descriptions have been considered in the literature, viz.

$$w = \sigma_x = v = 0$$
$$w = \sigma_x = \tau_{xy} = 0$$

for $x = $ constant edges, and similarly along $y = $ constant edges; the former can be identified to be the shear-diaphragm type described earlier, and the latter is a softer boundary condition where the in-plane tangential displacement is freely permitted. The solutions for these cases were obtained by Srinivas et al. [21] and Srinivas and Rao,[22] respectively.

A comparison of such results for a square plate with those of CPT reveals[23] that the shear forces at the shear-diaphragm-type edges, as obtained by integration of τ_{xz} and τ_{yz} through the thickness, coincide identically with Q_x and Q_y distributions of CPT for all span-to-thickness ratios and do not indicate the presence of any corner reactions. However, for the softer edge condition, the shear forces, obtained once again by integrating the shear stresses through the thickness, are close to the V_x, V_y distributions of CPT in the central portion of the edges; near the corners, the shear forces display negative (or downward) distributions that increase in magnitude and become more localized as the thickness of the plate decreases, thus indicating a trend towards becoming a concentrated reaction.

Thus, it appears that both the alternative support reactions mentioned above—Q_x, Q_y alone and V_x, V_y along with R—are acceptable as correct CPT predictions for the simply supported plate. The former should be used when the actual edge condition

---

[20]T. Kant, E. Hinton, Mindlin plate analysis by segmentation method, ASCE Jl. of Engineering Mechanics, 109(2), 1983, 537–556;

G. Prathap, C. Ramesh Babu, Accurate force evaluation with a simple, bilinear, plate bending element, Computers and Structures, 25(2), 1987, 259–270.

[21]S. Srinivas, A. K. Rao, C. V. Joga Rao, Flexure of simply supported thick homogeneous and laminated rectangular plates, ZAMM – Jl. of Applied Mathematics and Mechanics, 49(8), 1969, 449–458.

[22]S. Srinivas, A. K. Rao, Flexure of thick rectangular plates, ASME Jl. of Applied Mechanics, 40, 1973, 298–299.

[23]K. Rajaiah, A. K. Rao, A note on the reaction at simply supported square corners, ASME Jl. of Applied Mechanics, 45(1), 1978, 217–218.

can be specified as of the shear-diaphragm type, an example being the case of a portion of a large orthogonally stiffened plate bounded by open cross section stiffeners; such stiffeners have very low torsional rigidity so that the normal slope is freely permitted, but fairly high bending rigidity so that the in-plane tangential displacement, for all points on the edge surface including those away from the neutral mid-plane, has to be taken as fully restrained. For cases similar to a plate supported by a frame with the corners anchored down, the latter set of support reactions including the corner reactions is more appropriate.

10.6 Plates Under Thermal Loads

Three-dimensional elasticity solutions, similar to those described above for mechanical loading, may also be obtained for thermally loaded rectangular plates and laminates. For example, homogeneous isotropic and specially orthotropic plates with all edges supported by shear diaphragms and subjected to a temperature change $T(x,y,z)$ that is antisymmetric about the mid-plane may be analysed by expanding T as

$$T = \sum_m \sum_n T_{mn}(z) \sin \frac{m\pi x}{a} \sin \frac{n\pi y}{b} \qquad (10.33)$$

and seeking a solution for the displacements in the form of Eq. (10.24); simultaneously applied mechanical loading can readily be included in this analysis as described earlier. This methodology is also applicable for orthotropic cross-ply laminates.[24] Instead of an assumed variation of the temperature, one can also solve for it using the heat conduction equation when the temperatures on the six faces of the plate are specified, and then use it for subsequent stress analysis.[25]

Such solutions indicate that the classical plate theory is quite accurate for isotropic plates and laminates under thermal loading even when the thickness ratio is large; for orthotropic laminates, however, the accuracy deteriorates significantly when the coefficient of thermal expansion varies widely between the fibre (L) and transverse (T) directions. It should be noted that the ratio of α_T to α_L for certain unidirectional composite materials such as graphite-epoxy may be as high as 1000 and CPT errors are unacceptably large in such cases unless the plate is very thin. With regard to the corresponding displacement variations through the thickness, the in-plane displacements display the characteristic zigzag variation as described earlier with reference to mechanically loaded plates, while the transverse displacement, found to be nearly constant earlier, varies significantly because of significant thermal expansion in the unreinforced thickness direction.

---

[24] K. Bhaskar, T. K. Varadan, J. S. M. Ali, Thermoelastic solutions for orthotropic and anisotropic composite laminates, Composites Part B, 27B, 1996, 415–420.

[25] See, for example: V. B. Tungikar, K. M. Rao, Three-dimensional exact solution of thermal stresses in rectangular composite laminate, Composite Structures, 27, 1994, 419–430.

Summary

Some three-dimensional elasticity solutions for rectangular plates have been discussed. These solutions are useful for understanding the limitations of CPT and the actual behaviour of a plate that does not really fall within the purview of CPT. It has been shown that the CPT errors depend not only on the thickness ratio, but also several other factors such as the in-plane Young's modulus to transverse shear modulus ratio, the relative stiffness of one layer with respect to another, the nature of loading and boundary conditions. Another important conclusion is that a boundary layer region involving steep stress gradients exists near the edges for some support conditions and that this can be captured correctly only by using the three-dimensional elasticity approach.

Chapter 11
Shear Deformation Theories

In the previous chapter, the inadequacy of the classical plate theory for certain problems was explained based on comparison with three-dimensional elasticity solutions. Among the two non-classical effects, usually termed the *shear deformation effect* due to finite transverse shear rigidity and associated non-zero transverse shear strains, and the *thickness-stretch effect* due to Poisson contraction or elongation and finite transverse Young's modulus, the former has been shown to be more important; in fact, the latter has been shown to be negligibly small even for thick plates except in the vicinity of localized loads. While the three-dimensional approach is rigorous and accurate, it cannot be applied to every plate problem as a matter of routine because analytical solutions are possible only for a small number of cases and numerical solutions based on finite difference or finite element schemes tend to be quite demanding in terms of computer memory and execution time and also often exhibit convergence difficulties. Thus, there is a need for better two-dimensional plate theories based on hypotheses that are less restrictive and more realistic than Kirchhoff–Poisson hypothesis. These theories are generically referred to as shear deformation theories though some of them may also account for thickness stretch. Some such theories are discussed here starting with the simplest alternative to CPT.

11.1 First-Order Shear Deformation Theory (FSDT)

Confining attention to homogeneous plates subjected to transverse loads alone, it is easy to visualize the odd and even nature of the displacement functions with respect to z and to realize thereby that the transverse shear strains γ_{xz} and γ_{yz} are even functions. Thus, while hypothesizing their variation, the simplest possibility, next

© The Author(s) 2021
K. Bhaskar and T. K. Varadan, *Plates*,
https://doi.org/10.1007/978-3-030-69424-1_11

Fig. 11.1 Deformation as
per Mindlin's hypothesis

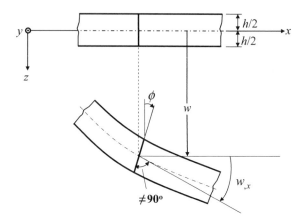

to taking them as negligibly small as in CPT, is to take them as constant through the thickness of the plate. The resulting theory is referred to as the *moderately thick plate theory*, or the *first-order theory* (because only the first-order z-terms occur in the in-plane displacement expansions as explained below), or Reissner–Hencky–Bollé–Mindlin theory after the four investigators who proposed it independently during 1945–51. We shall refer to it as FSDT.

In terms of the kinematics of deformation, one can state the fundamental assumption of this theory as follows:

Normals to the mid-plane of the undeformed plate suffer no change in length and remain straight, but not necessarily normal to the deformed mid-surface.

This statement is often referred to as *Mindlin's hypothesis* .

Thus the deformation is as shown in Fig. 11.1 which reveals that the rotation of the normal is different from the slope of the mid-surface at any point. Hence, the rotations $\phi\,(x, y)$ and $\psi\,(x, y)$ about the y and x axes, respectively, are independent variables in addition to $w\,(x, y)$.

The displacement field can hence be written as

$$
\begin{aligned}
u &= -z\phi \\
v &= -z\psi \\
w &= w(x, y)
\end{aligned}
\tag{11.1}
$$

where the third equation remains the same as for CPT because transverse normal strain is neglected here too.

The strain field is obtained as

$$
\begin{aligned}
\varepsilon_x &= u_{,x} = -z\phi_{,x} \\
\varepsilon_y &= v_{,y} = -z\psi_{,y} \\
\gamma_{xy} &= u_{,y} + v_{,x} = -z(\phi_{,y} + \psi_{,x})
\end{aligned}
\tag{11.2a}
$$

$$\gamma_{xz} = u_{,z} + w_{,x} = w_{,x} - \phi$$

$$\gamma_{yz} = v_{,z} + w_{,y} = w_{,y} - \psi \tag{11.2b}$$

The in-plane strains are linear in z as in CPT, while the transverse shear strains, neglected in CPT, turn out to be non-zero and constant through the thickness.

The plane-stress reduced constitutive law is taken to be applicable here to relate the in-plane stresses and strains. It is given by

$$\sigma_x = \frac{E}{(1 - \mu^2)} (\varepsilon_x + \mu\varepsilon_y)$$

$$\sigma_y = \frac{E}{(1 - \mu^2)} (\varepsilon_y + \mu\varepsilon_x)$$

$$\tau_{xy} = G\gamma_{xy} \tag{11.3a}$$

This is used along with

$$(\tau_{yz}, \tau_{xz}) = G(\gamma_{yz}, \gamma_{xz}) \tag{11.3b}$$

The resulting stress field is

$$\sigma_x = \frac{-Ez}{(1 - \mu^2)} (\phi_{,x} + \mu\psi_{,y})$$

$$\sigma_y = \frac{-Ez}{(1 - \mu^2)} (\mu\phi_{,x} + \psi_{,y})$$

$$\tau_{xy} = -Gz(\phi_{,y} + \psi_{,x}) \tag{11.4a}$$

$$\tau_{xz} = G(w_{,x} - \phi)$$

$$\tau_{yz} = G(w_{,y} - \psi) \tag{11.4b}$$

It should be noted that the in-plane stress field is linear through the thickness and anti-symmetric with respect to the mid-plane, which continues to be the neutral surface. Hence, the corresponding stress resultants are bending and twisting moments, defined as in CPT, as

$$(M_x, M_y, M_{xy}) = \int_{-h/2}^{h/2} (\sigma_x, \sigma_y, \tau_{xy}) z \, dz \tag{11.5}$$

The transverse shear stresses τ_{xz} and τ_{yz} are constant through the thickness and manifest themselves as the shear forces Q_x and Q_y given by

$$(Q_x, Q_y) = \int_{-h/2}^{h/2} (\tau_{xz}, \tau_{yz})\,dz \tag{11.6}$$

However, the above definition needs to be modified in view of the fact that a constant shear stress distribution through the thickness is unrealistic; this is so because the lateral surfaces of the plate are free of shear.

If one attempts to overcome this unrealistic variation by including higher-order z-terms in the in-plane displacement field, one ends up with a more complicated *higher-order theory*. Instead, to retain the simplicity characterized by just first-order z-terms, what is done is to introduce a correction factor, called the *shear correction factor* k^2, in Eq. (11.6) such that the resulting shear strain energy U_s is close to that of three-dimensional analysis. The modified equation is

$$(Q_x, Q_y) = k^2 \int_{-h/2}^{h/2} (\tau_{xz}, \tau_{yz})\,dz \tag{11.7}$$

For isotropic homogeneous plates, the most commonly used values for k^2 are 5/6 and $\pi^2/12$, as originally proposed by Reissner and Mindlin, respectively.

The stress resultants can be expressed in terms of the generalized displacements ϕ, ψ and w as

$$M_x = -D(\phi_{,x} + \mu\psi_{,y})$$
$$M_y = -D(\psi_{,y} + \mu\phi_{,x})$$
$$M_{xy} = -D\left(\frac{1-\mu}{2}\right)(\phi_{,y} + \psi_{,x}) \tag{11.8a}$$

$$Q_x = k^2 Gh(w_{,x} - \phi)$$
$$Q_y = k^2 Gh(w_{,y} - \psi) \tag{11.8b}$$

Since the stress resultants are of the same nature as in CPT, the derivation of the equilibrium equations in terms of them remains the same (see Sect. 2.3). For the sake of completeness, these equations are reproduced below.

$$Q_{x,x} + Q_{y,y} + q = 0$$
$$M_{x,x} + M_{xy,y} - Q_x = 0$$
$$M_{xy,x} + M_{y,y} - Q_y = 0 \tag{11.9}$$

The corresponding governing equations are

$$k^2 Gh(w_{,xx} + w_{,yy} - \phi_{,x} - \psi_{,y}) + q = 0$$

$$D\left[\phi_{,xx} + \left(\frac{1-\mu}{2}\right)\phi_{,yy} + \left(\frac{1+\mu}{2}\right)\psi_{,xy}\right] + k^2 Gh(w_{,x} - \phi) = 0$$

$$D\left[\psi_{,yy} + \left(\frac{1-\mu}{2}\right)\psi_{,xx} + \left(\frac{1+\mu}{2}\right)\phi_{,xy}\right] + k^2 Gh(w_{,y} - \psi) = 0 \qquad (11.10)$$

This system of differential equations is of sixth order, and hence three boundary conditions can be satisfied at any edge. The classical boundary conditions are.

(a) **Simply supported edge**

At $x = $ constant : $w = M_x = 0$ with either $M_{xy} = 0$ or $\psi = 0$

At $y = $ constant : $w = M_y = 0$ with either $M_{xy} = 0$ or $\phi = 0$ $\qquad (11.11)$

Hence, one can distinguish between two types of simple supports—one for which the tangential displacement along the edge is completely restrained (corresponding to the shear-diaphragm-type support), and the other for which it is completely unrestrained and hence the twisting moment M_{xy} is zero. Thus, the boundary constraint description is better here than in CPT.

(b) **Clamped edge**

At $x = $ constant : $w = \phi = 0$ with either $M_{xy} = 0$ or $\psi = 0$

At $y = $ constant : $w = \psi = 0$ with either $M_{xy} = 0$ or $\phi = 0$ $\qquad (11.12)$

where, once again, one can distinguish between the cases of totally restrained and totally unrestrained along-the-edge displacements.

(c) **Free edge**

All the stress resultants corresponding to a free edge are equated to zero, as given by

$$Q_x = M_x = M_{xy} = 0 \text{ at } x = \text{constant}$$

$$Q_y = M_y = M_{xy} = 0 \text{ at } y = \text{constant} \qquad (11.13)$$

which coincide with Poisson's boundary conditions of Sect. 2.5. Thus, the difficulty faced in CPT regarding the satisfaction of three conditions and the necessity of the fictitious Kirchhoff's shear force are obviated here.

Equation (11.10) along with the appropriate boundary conditions have to be satisfied to obtain the solution for the problem. Once ϕ, ψ and w are determined, the strains and stresses can be calculated at any point of the plate. However, the transverse shear stresses obtained by the stress–strain law turn out to be constant through the thickness and hence not acceptable because they violate the shear-free lateral surface conditions. This difficulty can be overcome by obtaining statically equivalent estimates for them as was done in CPT (Sect. 2.6).

The extension of this theory to anisotropic homogeneous plates, with a plane of symmetry parallel to the mid-plane (see Sect. 9.1.1), is straightforward—the only difference is that the anisotropic constitutive law [Eq. (9.8)] is used in place of Eq. (11.3). It should be noted that there is coupling between the two transverse shears for such plates as given by[1]

$$\begin{pmatrix} \tau_{yz} \\ \tau_{xz} \end{pmatrix} = \begin{bmatrix} \overline{Q}_{44} & \overline{Q}_{45} \\ \overline{Q}_{45} & \overline{Q}_{55} \end{bmatrix} \begin{pmatrix} \gamma_{yz} \\ \gamma_{xz} \end{pmatrix} \tag{11.14}$$

The above changes will result in corresponding changes in the generalized force–displacement relations, which can be written as

$$\begin{pmatrix} M_x \\ M_y \\ M_{xy} \end{pmatrix} = \begin{bmatrix} D_{11} & D_{12} & D_{16} \\ & D_{22} & D_{26} \\ \text{sym.} & & D_{66} \end{bmatrix} \begin{pmatrix} \phi_{,x} \\ \psi_{,y} \\ \phi_{,y} + \psi_{,x} \end{pmatrix}$$

$$\begin{pmatrix} Q_y \\ Q_x \end{pmatrix} = \begin{bmatrix} k_1^2 A_{44} & k_2^2 A_{45} \\ \text{sym.} & k_3^2 A_{55} \end{bmatrix} \begin{pmatrix} w_{,y} - \psi \\ w_{,x} - \phi \end{pmatrix} \tag{11.15}$$

wherein one should note that there is an option to choose different shear correction factors for different stiffness coefficients.

The following example serves to illustrate the improved accuracy of FSDT as compared to that of CPT for a homogeneous isotropic plate.

Example 11.1: Cylindrical Bending of a Moderately Thick Plate Strip

The problem of an infinite plate strip, simply supported at the two longitudinal edges and subjected to transverse sinusoidal load—for which elasticity and CPT solutions were obtained in Chaps. 1 and 3, respectively—is considered here. Putting $\psi = (\ )_{,y} = 0$ in the governing equations and the boundary conditions, one obtains

$$k^2 Gh(w_{,xx} - \phi_{,x}) + q_o \sin \frac{\pi x}{a} = 0$$

$$D\phi_{,xx} + k^2 Gh(w_{,x} - \phi) = 0$$

along with

$$w = M_x = 0$$

---

[1]If this pertains to a generally orthotropic layer of fibre orientation θ, the transformed stiffness coefficients can be expressed in terms of those with respect to the principal material axes [see Eq. (9.1)] as.

$\overline{Q}_{44} = c^2 C_{44} + s^2 C_{55}; \quad \overline{Q}_{55} = c^2 C_{55} + s^2 C_{44}; \quad \overline{Q}_{45} = cs(C_{55} - C_{44})$

where $c = \cos \theta$, $s = \sin \theta$.

$$\text{i.e.} \quad w = \phi_{,x} = 0 \quad \text{at } x = 0, a.$$

A Navier-type solution satisfying the boundary conditions can be sought as

$$w = W \sin \frac{\pi x}{a}; \qquad \phi = P \cos \frac{\pi x}{a}$$

Substitution of this solution in the governing equations reduces them to two algebraic equations in W and P, which can be solved to yield

$$W = \frac{q_o a^4}{\pi^4 D} + \frac{q_o a^2}{k^2 \pi^2 Gh}; \qquad P = \frac{q_o a^3}{\pi^3 D}$$

where the first term of the deflection can be identified to be the bending deflection obtained using CPT and the second term is the shear deflection.

Taking the ratio of the two components, one gets

$$\frac{W_{shear}}{W_{bending}} = \frac{\pi^2}{12(1 - \mu^2)k^2} \left(\frac{E}{G}\right)\left(\frac{h}{a}\right)^2$$

which is quite similar to that obtained for the cantilevered strip using the elasticity approach in Chap. 10.

The bending stress and the strain energy per unit length of the infinite strip are calculated using

$$\sigma_x = \frac{-Ez}{(1 - \mu^2)}\phi_{,x}$$

$$U_e = \frac{1}{2} \iint (\sigma_x \varepsilon_x + k^2 \tau_{xz}\gamma_{xz})dx\,dz$$

$$= \frac{1}{2} \int [D\phi_{,x}^2 + k^2 Gh(w_{,x} - \phi)^2]dx$$

At this point, it should be noted that the present problem of cylindrical bending is statically determinate, and the bending stress distribution through the thickness at any section should be statically equivalent to the bending moment M_x at that section which is readily known from the given sinusoidal transverse load. Thus, *for the present problem alone*, any theory which is based on a linear bending stress distribution should lead to the same values for σ_x, as given by

$$\sigma_x = \frac{12z M_x}{h^3}$$

Table 11.1 Results of FSDT

a/h	$U_e{}^*$			$w^*(a/2, 0)$		
	FSDT	**Elasticity**	**Error (%)**	**FSDT**	**Elasticity**	**Error (%)**
5	0.0312	0.0304	2.6	0.1247	0.123	1.1
10	0.0288	0.0286	0.8	0.1153	0.115	0.3
25	0.0281	0.0281	≈ 0	0.1126	0.113	≈ 0
50	0.0281	0.0280	≈ 0	0.1122	0.112	≈ 0
CPT	0.0280			0.112		

Note: μ is taken to be 0.3, and k^2 to be 5/6

$\text{Error} = \frac{\text{(FSDT result} - \text{Elasticity result)}}{\text{Elasticity result}}$, shown as '$\approx 0$' when $< 0.1\%$

and this is true for both CPT and FSDT. Since the statically equivalent estimates of the transverse shear stress τ_{xz} are obtained from the bending stress distribution, they will also be same for this particular problem by both the theories.

Numerical results for the maximum deflection and the total strain energy are compared in Table 11.1 for various values of a/h; the elasticity and CPT estimates are also included for comparison. The non-dimensionalization is the same as used earlier in Table 3.1.

From this Table, it can be seen that FSDT yields excellent results for $a/h \geq 10$ for the problem under consideration. A comparison of the error percentages given here with those of CPT in Table 3.1 enables one to appreciate the improved accuracy of FSDT.

The first-order theory was extended to arbitrarily laminated anisotropic plates by Yang, Norris and Stavsky in 1966, and this version, as documented with a few minor changes by Whitney and Pagano,[2] is also referred to as *YNS Theory*; in this theory, mid-plane displacements u_o and v_o are included to account for bending–stretching coupling, and correspondingly there are two additional governing equations.

In their paper, Whitney and Pagano showed that exact Navier-type solutions are possible for this theory, as was the case with Classical Laminated Plate Theory (see Sect. 9.2), for two types of unsymmetric laminates—the cross-ply plate with shear-diaphragm-type edges and the anti-symmetric angle-ply plate with another type of simple supports analogous to those described by Eq. (9.33).

For problems of dynamics and stability, appropriate terms have to be added to the governing equations; while this step is exactly identical to that in CPT for the stability problem, additional inertia terms besides the transverse inertia, neglected earlier for the thin plate but not negligible for the present moderately thick config-uration, come into picture for the dynamic problem. Considering a homogeneous plate or a symmetric laminate, these additional terms correspond to inertia moments due to rotations of the normal about the y and x axes as given by $-\int \rho z^2 \mathrm{d}z\, \phi_{,tt}$ and

---

[2]J. M. Whitney, N. J. Pagano, Shear deformation in heterogeneous anisotropic plates. ASME Jl. of Applied Mechanics, 37, 1970, 1031–1036.

$-\int \rho z^2 dz\, \psi_{,tt}$; these are added to the left-hand side of the appropriate moment equilibrium equations [Eqs. (11.9)]. For an unsymmetric laminate, in-plane inertia effects may also be considered by including the terms $-\bar{\rho}u_{o,tt}$, $-\bar{\rho}v_{o,tt}$ in the corresponding in-plane equilibrium equations, but their effect is usually negligibly small.

Levy-type solutions[3] are also possible for moderately thick plates with two opposite edges simply supported; once again, as with classical theory solutions, elegant double series counterparts can be derived.[4]

The accuracy and other advantages of FSDT have been examined with reference to a large number of problems in the literature, and the many conclusions may be summarized as follows:

(a)   FSDT yields more accurate estimates for deflections, natural frequencies and buckling loads than CPT and is certainly useful for this purpose even for fairly thick plates.

(b)   For the associated stresses, however, improved accuracy is not guaranteed as some of the stress components may be better and some worse than the CPT counterparts.

(c)   The accuracy of the results varies significantly with the shear correction factor employed, and the best value for this factor depends on the loading, boundary conditions and geometrical details such as ply sequence and also on whether the analysis is for deflections or vibration frequencies or buckling loads. Though a number of methods have been proposed for determining this factor,[5] none is them is unanimously accepted as the best or most convenient.

(d)   FSDT is more convenient for finite element implementation than CPT because it requires just C^o continuity (i.e. continuity of the displacement variables w, ϕ, ψ alone and not their derivatives) as against C^1 continuity (i.e. continuity of w as well as its first derivatives $w_{,x}$ and $w_{,y}$) for CPT. However, while the results generated by an analytical solution of the governing equations display a correct coincidence with those of CPT when the plate is very thin, the finite element results for such a thin plate may exhibit numerical difficulties due a phenomenon referred to as *shear locking*[6] unless this is carefully addressed during the finite element formulation.

---

[3]J. N. Reddy, A. A. Khdeir, L. Librescu, Levy type solutions for symmetrically laminated rectangular plates using first-order shear deformation theory, ASME Jl. of Applied Mechanics, 54, 1987, 740–742.

[4]S. Kshirsagar, K. Bhaskar, Free vibration and stability analysis of orthotropic shear-deformable plates using untruncated infinite series superposition method, Thin-walled Structures, 47, 2009, 403–411.

[5]See, for example:

A. K. Noor, W. S. Burton, J. M. Peters, Predictor–corrector procedure for stress and free vibration analyses of multilayered composite plates and shells, Computer Methods in Applied Mechanics and Engineering, 82, 1990, 341–364;

P. F. Pai, A new look at shear correction factors and warping functions of anisotropic laminates, International Jl. of Solids and Structures, 32, 1995, 2295–2313.

[6]See, for example: D. Briassoulis, On the basics of the shear locking problem of C^o isoparametric plate elements, Computers and Structures, 33, 1989, 169–185.

Fig. 11.2 Warping of the
normal

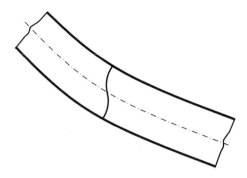

(e)   With reference to the relative importance of shear deformation and rotary
       inertia, the former has been found to be more significant.[7]

11.2 Higher-Order Theories

As mentioned earlier, theories which include higher-order z-terms beyond the linear
term in the u,v displacement fields are referred to as *higher-order theories* . The
main motivation for these more complicated theories is to eliminate the arbitrary shear
correction factor and to obtain better estimates of stresses and in-plane displacements
by accounting for the warping of the normals (Fig. 11.2).

The simplest higher-order theory for a homogenous plate is based on the following
displacement field:

$$u(x, y, z) = zu_1(x, y) + z^3 u_3(x, y)$$
$$v(x, y, z) = zv_1(x, y) + z^3 v_3(x, y)$$
$$w(x, y, z) = w(x, y) \tag{11.16}$$

which corresponds to approximation of the deformed shape of the normal as a cubic
curve, and complete neglect of thickness stretch. Thus, the in-plane stresses are also
cubic in z, while the transverse shear stresses are quadratic. The zero shear conditions
on the top and bottom surfaces of the plate are not enforced a priori, but are expected
to be nearly satisfied by the final solution.

Alternatively, one can decide to enforce these lateral zero shear conditions
explicitly in the beginning itself, to yield

$$\gamma_{xz} = u_1 + w_{,x} + 3z^2 u_3 = 0 \quad \text{at } z = \pm h/2$$

---

[7]C. W. Bert, M. Malik, On the relative effects of transverse shear deformation and rotary inertia on
the free vibration of symmetric cross-ply laminated plates, Jl. of Sound and Vibration, 193, 1996,
927–933.

$$\text{i.e.} \qquad u_3 = -\frac{4}{3h^2}(u_1 + w_{,x})$$

$$\text{i.e.} \qquad \gamma_{xz} = (u_1 + w_{,x})\left(1 - \frac{4z^2}{h^2}\right) \tag{11.17}$$

and a similar expression for γ_{yz}. Thus, the resulting theory is one corresponding to a parabolic distribution of the transverse shear strains which is more realistic than the uniform distribution assumed in FSDT. This kind of a *parabolic shear strain theory* was first proposed by Vlasov in 1957[8] and also independently by several others later, the inspiration being provided by CPT itself wherein the transverse shear stresses display such a variation when obtained as statically equivalent estimates using the three-dimensional equilibrium equations.

Similar to the above two theories, one can think of more complicated theories with quintic z-terms, etc., in the u, v expansions, or their counterparts satisfying the zero shear conditions on the lateral surfaces. All such theories can be straightaway extended for symmetric laminates and further for unsymmetric laminates by introducing even z-terms (often just the constant u_o and v_o terms) in the u, v expansions. Instead of simple polynomial terms, one can employ orthogonal polynomials or even trigonometric functions of z.

Theories for laminates as described above, based on an approximation for the displacements over the entire thickness of the laminate, are referred to as *equivalent single layer theories*. These theories employ continuous smoothly varying functions and hence cannot capture the piecewise, zigzag nature of the displacements as revealed by the elasticity solutions described in Chap. 10. To address this deficiency, one can employ approximate displacement fields (linear, quadratic, etc., or some other variations) for each layer with continuity imposed at the interfaces to result in what are called *discrete layer theories*. An important difference between equivalent single layer theories and discrete layer theories is that the number of displacement variables (functions of x, y) remains independent of the number of layers for the former while it grows in direct proportion with the number of layers for the latter; thus, discrete layer theories are in general more computationally intensive, but also more accurate. Discrete layer theories may further be classified on the basis of whether the interface stress continuity conditions are enforced a priori or not.

An ingenious approach of accounting for sudden changes of slope of the displacement variations at the interfaces is to introduce a so-called *zigzag function* in addition to globally continuous functions in the displacement expansions of an equivalent single layer theory; a simple example of a zigzag function is one[9] that varies linearly within each layer and takes values of 1 and -1 at the successive interfaces

---

[8]See: G. Jemielita, On kinematical assumptions of refined theories of plates: A survey, ASME Jl. of Applied Mechanics, 57, 1990, 1088–1091.

[9]H. Murakami, Laminated composite plate theory with improved in-plane responses, ASME Jl. of Applied Mechanics, 53, 1986, 661–666.

Fig. 11.3 Zigzag function

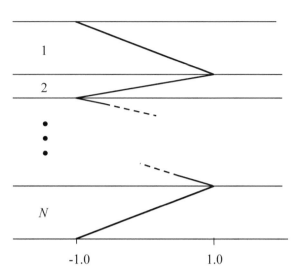

(Fig. 11.3). Such *zigzag theories*[10] are in general more accurate than conventional equivalent single layer theories of the same complexity and a convenient alternative to the costly discrete layer theories.

Another possibility is to start with a simple equivalent single layer theory such as the classical theory to obtain the in-plane stress field and the statically equivalent transverse stresses; the transverse strains are then obtained using the constitutive law, and they are integrated to obtain an improved displacement field. This can be repeated to obtain a hierarchy of more accurate theories—the procedure itself being referred to as iterative modelling.[11]

With laminates with a very large number of layers, it is also possible to model a group of layers as an equivalent single layer by postulating a displacement or stress variation for the total thickness of this chosen group, and to model the remaining individual layers using a discrete layer approach, with appropriate conditions to ensure continuity at the interface between the two sets of layers. This is referred to as a global–local model.[12]

While developing a higher-order theory, one should be careful to derive the equations using the variational approach starting with the principle of virtual work or the principle of minimum potential energy because one is likely to miss out certain terms in the equations in the conventional vectorial approach of summing forces and

---

[10]E. Carrera, Historical review of zigzag theories for multilayered plates and shells, Applied Mechanics Reviews, 56, 2003, 287–308.

[11]K. Vijayakumar, A.V. Krishna Murty, Iterative modeling for stress analysis of composite laminates, Composite Structures, 32, 1988, 165–181.

[12]N. J. Pagano, S. R. Soni, Global–local laminate variational model, International Jl. of Solids and Structures, 19, 1983, 207–228.

moments. This has been pointed out clearly by Reddy[13] with respect to the parabolic shear strain theory.

Another important feature of higher-order theories is that one encounters higher-order stress resultants such as $\int \sigma_x z^3 dz$ and $\int \tau_{xz} z^2 dz$ which do not readily admit of a physical visualization. Further, the boundary conditions are specified in terms of these resultants and their corresponding higher-order displacement terms [such as u_3 in Eq. (11.16)]; this can be of advantage in properly describing a practical restraint, or irksome when the nature of the restraint is not very clear.

The theories described above are all based on assumed displacement fields. Alternatively, since the transverse stresses τ_{xz}, τ_{yz} and σ_z are continuous across the interfaces of a laminate, one can conveniently postulate their distributions along with those of the displacements leading a mixed variational formulation.[14] There are also other methods of developing higher-order theories such as a systematic reduction of the three-dimensional elasticity equations to recursive sets of two-dimensional equations using asymptotic integration; for a discussion of all such methods, one is referred to appropriate review articles[15] that periodically survey the fast-growing field of plate theories and conveniently categorize them.

Many attempts have been made to assess the accuracy of popular higher-order theories against three-dimensional solutions.[16] It is not very easy to summarize the conclusions from all these studies because each such study was specific to a particular problem and a corresponding set of results and was also confined to a small selected set of theories. However, in order to motivate interest in such comparison studies and to give an overall perspective of the importance of shear deformation theories, we shall present below several salient conclusions which are true in general, but not without exceptions.

---

[13] J. N. Reddy, A simple higher-order theory for laminated composite plates, ASME Jl. of Applied Mechanics, 51, 1984, 745–752.

[14] E. Reissner, On a certain mixed variational theorem and a proposed application, International Jl. of Numerical Methods in Engineering, 20, 1984, 1366–1368.

[15] See, for example:

Y. M. Ghugal, R.P. Shimpi, A review of refined shear deformation theories of isotropic and anisotropic laminated plates, Jl. of Reinforced Plastics and Composites, 21, 2002, 775–813;

H. Altenbach, Theories for laminated and sandwich plates: a review, Mechanics of Composite Materials, 34(3), 1998, 243–252.

[16] See, for example:

A. K. Noor, W. S. Burton, Assessment of shear deformation theories for multilayered composite plates, Applied Mechanics Reviews, 42, 1989, 1–13;

M. R. Chitnisa, Y. M. Desai, A. H. Shahb, T. Kant, Comparisons of displacement-based theories for waves and vibrations in laminated and sandwich composite plates, Jl. of Sound and Vibration, 263, 2003, 617–642;

E. Carrera, A. Ciuffreda, Bending of composites and sandwich plates subjected to localized lateral loadings: a comparison of various theories, Computers and Structures, 68, 2005, 185–202.

(a)   With a polynomial displacement expansion, while the cubic z-terms denote a significant improvement over just the linear term for the analysis of homogeneous plates, the additional benefit due to higher-order terms is very marginal.

(b)   For laminates, the inclusion of the zigzag nature of the in-plane displacement field with sharp changes of slopes at the interfaces is important, and this can be adequately achieved by using a zigzag function within an equivalent single layer approach, thus obviating the more computationally intensive discrete layer approach.[17]

(c)   The accurate estimation of transverse shear stresses requires integration of the three-dimensional equilibrium equations irrespective of whether the lateral boundary conditions and interface conditions are satisfied a priori or not. From this viewpoint, it appears preferable not to satisfy such conditions so that the displacement field remains simple, and also one that requires just C^o continuity for finite element implementation. It has also been shown,[18] with reference to the theory based on Eq. (11.16), that an attempt to enforce the lateral zero shear conditions a priori results in slightly worse predictions of deflections and in-plane stresses instead of better ones.

(d)   The thickness-stretch effect need not be included except for stress analysis near local loads and for problems involving thermal loading where the coefficient of expansion in the unreinforced thickness direction can be several orders of magnitude larger than that along the fibre direction.[19] When the thickness-stretch effect is included, the assumption of the plane-stress constitutive law is no more necessary.

(e)   While establishing limits of applicability of any two-dimensional theory, it is necessary to consider more complicated cases (localized loads, clamped edge conditions) in addition to the commonly considered case of simply supported plates subjected to sinusoidal or uniform loading because this is perhaps the least critical test case in terms of the severity of non-classical effects.

(f)   For sandwich plates, the relative stiffness ratio of the face-sheets with respect to the core is an important parameter that affects the accuracy of two-dimensional theories, as has been pointed out with reference to the cantilevered strip in Chap. 10; for high values of this parameter, simple equivalent single layer theories fail to provide acceptable results, and one has to necessarily use a zigzag theory or the discrete layer approach.[20]

---

[17]T. K. Varadan, K. Bhaskar, Review of different laminate theories for the analysis of composites, Jl. of the Aeronautical Society of India, 49(4), 1997, 202–208.

[18]B. N. Pandya, T. Kant, Higher-order shear deformable theories for flexure of sandwich plates—finite element evaluations, International Jl. of Solids and Structures, 24, 1988, 1267–1286.

[19]J. S. M. Ali, K. Bhaskar, T. K. Varadan, A new theory for accurate thermal/mechanical flexural analysis of symmetrically laminated plates, Composite Structures, 45, 1999, 227–232.

[20]E. Carrera, S. Brischetto, A survey with numerical assessment of classical and refined theories for the analysis of sandwich plates, Applied Mechanics Reviews, 62, 2009, Paper No. 010803.

Summary The objective of this chapter is to introduce two-dimensional theories that are based on more realistic assumptions as compared to the classical thin plate theory. While the first-order theory has been put forth in full detail and its accuracy demonstrated with respect to a simple example, higher-order theories have been discussed only briefly with appropriate references such that the reader gets an idea of the wide variety of approaches employed for accurate analysis of thick plate structures. However, important conclusions drawn from several comparison studies have been highlighted.

Chapter 12
Variable Thickness Plates

While plates of uniform thickness are commonly used, variable thickness plates are also sometimes encountered, as turbine discs and rocket and missile fins for example. Variable thickness plates also form an important part of structural optimization studies wherein one seeks to obtain the maximum fundamental frequency for a given total weight, or a fully stressed configuration wherein the material is effectively utilized at every point of the plate domain, or some such benefit. The analysis of plates of non-uniform thickness is much more complicated than that of uniform plates because variable coefficients occur in the governing equations. The purpose of this chapter is to cite some special cases that are amenable to exact analysis, with a brief mention of the applicable methodology.

12.1 Stepped Versus Smooth Thickness Variation

With respect to the analysis of variable thickness plates, it is first necessary to distinguish between stepped plates and plates with continuously varying thickness and to note that the former case is easily analysed by splitting the total plate domain into sub-domains of uniform thickness and by imposing appropriate continuity conditions along the intersections.[1]

The analysis of plates with continuous thickness variation is more complicated and discussed in this chapter. In this context, it is appropriate to note that variable thickness is modelled within the purview of CPT in terms of a variable flexural rigidity D, and thus the solutions discussed here are also applicable for functionally graded

---

[1] See, for example: Y. Xiang, C. M. Wang, Exact vibration and buckling solutions for stepped rectangular plates, Jl. of Sound and Vibration, 250, 2002, 503–517.

© The Author(s) 2021
K. Bhaskar and T. K. Varadan, *Plates*,
https://doi.org/10.1007/978-3-030-69424-1_12

plates of uniform thickness with appropriate gradations of the material properties (E, μ, ρ) in the in-plane directions.

12.2   Solutions for Rectangular Plates

The moment–curvature relations of Eq. (2.13) are applicable for the variable thickness plate as well, except that D is now a function of the coordinates x, y. Substitution of these in the equilibrium equation (Eq. 2.17) reduces it to

$$D\nabla^4 w + 2D_{,x}(\nabla^2 w)_{,x} + 2D_{,y}(\nabla^2 w)_{,y} + \nabla^2 D\ \nabla^2 w$$
$$- (1-\mu)(D_{,yy}w_{,xx} - 2D_{,xy}w_{,xy} + D_{,xx}w_{,yy}) = q \qquad (12.1)$$

in lieu of the simple biharmonic equation for the uniform plate.

This equation, in all its generality, does not admit of rigorous solutions and can only be solved approximately using finite difference or perturbation or some other method; alternatively, one can employ an energy method such as the Rayleigh–Ritz method with the strain energy given by Eq. (2.42) wherein $D(x,y)$ has to be taken within the integral now.[2]

For special cases, exact solutions are possible; these are for plates simply supported on two opposite edges and with the thickness varying only in the direction parallel to these edges. Taking this to be the y-direction, the governing equation gets simplified to

$$D\nabla^4 w + 2D_{,y}(\nabla^2 w)_{,y} + D_{,yy}\ (w_{,yy} + \mu\, w_{,xx}) = q \qquad (12.2)$$

The simplest case[3] in this category of problems is when the load is uniform ($q = q_o$) and the flexural rigidity variation is exponential as given by

$$D = D_o e^{\alpha y} \qquad (12.3)$$

Then the governing equation reduces to one with constant coefficients as given by

$$\nabla^4 w + 2\alpha(\nabla^2 w)_{,y} + \alpha^2(w_{,yy} + \mu\, w_{,xx}) = \frac{q_o}{D_o}e^{-\alpha y} \qquad (12.4)$$

---

[2] See, for example: Y. K. Cheung, D. Zhou, The free vibrations of tapered rectangular plates using a new set of beam functions with the Rayleigh–Ritz method , Jl. of Sound and Vibration, 223, 1999, 703–722.

[3] H. D. Conway, A Levy-type solution for a rectangular plate of variable thickness, ASME Jl. of Applied Mechanics, 25, 1958, 297–298.

This is solved using a Levy-type methodology by seeking the deflection function as

$$w = e^{-\alpha y} w_1(x) + \sum_m W_m(y) \sin \frac{m \pi x}{a} \tag{12.5}$$

such that the first term satisfies the non-homogeneous governing equation, and $W_m(y)$ is found out using its homogeneous counterpart. This solution can be written in a generic form in terms of four constants for each m, and these constants are evaluated by imposing the conditions at the edges parallel to the x-axis, say, at $y = \pm b/2$. This exact solution is important because the assumed exponential variation almost coincides with the practically significant case of linear variation when the plate taper is small.

As shown by Mansfield,[4] the above approach is also valid when the load q is not constant but varies exponentially in the y-direction as

$$q = q(x) e^{\beta y} \tag{12.6}$$

In this case, the solution for w is sought as

$$w = e^{(\beta - \alpha)y} w_1(x) + \sum_m W_m(y) \sin \frac{m \pi x}{a} \tag{12.7}$$

A very similar approach is applicable for the problem[5] involving a linear taper and with the load taken to be proportional to the flexural rigidity; such a load variation, though unrealistic, is assumed because it leads to a closed-form solution in terms of exponential integrals.

The Levy-type approach is also applicable to the above category of problems with reference to frequency and stability analysis. For the free vibration problem, the load term in Eq. (12.2) is replaced by the inertia term—$\rho h w_{,tt}$ where ρ is the mass density. Seeking a general normal mode as

$$w(x, y) = W_m(y) \sin \frac{m \pi x}{a} \tag{12.8}$$

one gets, for the eigenvalue problem on hand, a homogeneous ordinary differential equation of the form

$$W_{m,yyyy} + f_3(y) W_{m,yyy} + f_2(y) W_{m,yy}$$
$$+ f_1(y) W_{m,y} + f_0(y) W_m = 0 \tag{12.9}$$

---

[4]E. H. Mansfield, On the analysis of elastic plates of variable thickness, Quarterly Jl. of Mechanics and Applied Mathematics, XV, Part 2, 1962, 167–192.

[5]S. P. Timoshenko, S. W. Krieger, Theory of Plates and Shells, McGraw-Hill, 1959, pp. 174–176.

Seeking a power series solution about an ordinary point $y = y_o$ where no singularities occur (i.e. where the functions $f_i(y)$ are all analytic), one assumes

$$W_m(y) = \sum_{n=0,1,2,\dots} C_n(y - y_o)^n \qquad (12.10)$$

After substitution of this in Eq. (12.9) and equating the coefficient of each power of $(y - y_o)$ to zero, one gets a set of recurrence relations yielding higher constants C_n in terms of the first four constants C_0 to C_3. By imposing the edge conditions at $y = \pm b/2$, and seeking a non-trivial solution, the frequency equation is obtained. This can be solved numerically, taking care to evaluate the power series accurately by taking a large number of terms.

The above procedure is valid for a general polynomial variation of the thickness in the y-direction. It is also valid for the problem of buckling due to compressive load $N_x(y)$ applied along the edges $x = 0,a$ with a polynomial variation along the edges; in this case one has $N_x w_{,xx}$ in place of q in the governing equation Eq. (12.2).

Such rigorous power series solutions have been obtained for the vibration frequencies of linearly tapered rectangular plates with combinations of simply supported, clamped and free edge conditions along $y = \pm b/2$ by Akiyama and Kuroda[6] and for elastically restrained edge conditions by Kobayashi and Sonoda.[7]

For the buckling problem, similar solutions have been obtained[8] for several boundary conditions at $y = \pm b/2$ for the case of a linear taper and linearly varying N_x. More recently,[9] this approach has been used to study plates with a linear variation of the Young's modulus, or with linear, quadratic and cubic variations of the thickness as well as the load.

12.3 Circular Plates

In this category, we shall confine attention to axisymmetric problems with the thickness varying in the radial direction alone, a case commonly encountered in practice.

The equilibrium equations in terms of the stress resultants are obtained by specializing Eq. (5.9) for axisymmetry as

[6]K. Akiyama, M. Kuroda, Fundamental frequencies of rectangular plates with linearly varying thickness, Jl. of Sound and Vibration, 205, 1997, 380–384.

[7]H. Kobayashi, K. Sonoda, Vibration and buckling of tapered rectangular plates with two opposite edges simply supported and the other two edges elastically restrained against rotation, Jl. of Sound and Vibration, 146, 1991, 323–337.

[8]H. Kobayashi, K. Sonoda, Bucklings of rectangular plates with tapered thickness, ASCE Jl. of Structural Engineering, 116, 1990, 1278–1289.

[9]M. Saeidifar, S. N. Sadeghi, M. R. Saviz, Analytical solution for the buckling of rectangular plates under uni-axial compression with variable thickness and elasticity modulus in the y-direction, I. Mech. E. Jl. of Mechanical Engineering Science, 224, 2010, 33–41.

$$Q_{r,r} + \frac{Q_r}{r} + q = 0 \tag{12.11}$$

$$M_{r,r} + \frac{(M_r - M_\theta)}{r} - Q_r = 0 \tag{12.12}$$

and these can be combined to yield

$$M_{r,rr} + \frac{2M_r}{r} - \frac{M_{\theta,r}}{r} + q = 0 \tag{12.13}$$

For solid circular plates and for annular plates with the inner boundary unsupported, one can determine $Q_r(r)$ from the given $q(r)$ by integration of Eq. (12.11) as

$$Q_r = -\frac{1}{r} \int_{r_{inner}}^{r} q r \, dr \tag{12.14}$$

and then use Eq. (12.12); in other cases, one has to use Eq. (12.13). For further discussion here, we shall assume $Q(r)$ to be known.

Using the moment–curvature relations of Eq. (5.8) without the θ-derivative terms and taking D to be a function of r, Eq. (12.12) can be rewritten as

$$D\left(\psi_{,rr} + \frac{\psi_{,r}}{r} - \frac{\psi}{r^2}\right) + D_{,r}\left(\psi_{,r} + \mu\frac{\psi}{r}\right) = Q_r \tag{12.15}$$

where ψ is the normal slope given by

$$\psi = -w_{,r} \tag{12.16}$$

While a general solution of Eq. (12.15) for arbitrary $D(r)$ is not possible without recourse to approximate numerical methods, several special cases are amenable to exact analysis.

The simplest of these is when the thickness variation is given by

$$h = h_o r^m \tag{12.17}$$

as applicable to, say, an annular plate. The exponent m may be positive or negative leading to an increase or decrease of thickness with r. With D and $D_{,r}$ given by

$$D = D_o r^{3m}; \quad D_{,r} = D_o r^{3m-1} = D/r \tag{12.18}$$

where D_o corresponds to h_o, Eq. (12.15) reduces to

$$\psi_{,rr} + \frac{2\psi_{,r}}{r} - \frac{(1-\mu)\psi}{r^2} = \frac{Q_r}{D_o r^{3m}} \tag{12.19}$$

This equation is of the Euler–Cauchy form and can be converted to an equation with constant coefficients by using the coordinate transformation

$$r = e^s \quad \text{or} \quad s = \ln r \tag{12.20}$$

and hence

$$\psi_{,r} = \psi_{,s} s_{,r} = \frac{\psi_{,s}}{r}; \qquad \psi_{,rr} = \frac{\psi_{,ss} - \psi_{,s}}{r^2} \tag{12.21}$$

Thus, Eq. (12.19) reduces to

$$\psi_{,ss} + \psi_{,s} - (1-\mu)\psi = \frac{Q_r}{D_o} e^{(2-3m)s} \tag{12.22}$$

which admits of a straightforward solution for a specified $Q_r(r)$.

The two undetermined constants are found out by imposing the conditions at the inner and outer boundaries, and the deflection itself is found out by integrating ψ with respect to r using Eq. (12.16). Such a solution for $m = 1$ was obtained by Conway in 1948 and used to generate results for the maximum deflection and maximum bending stress for several practical cases.

The other special cases admitting a straightforward solution are not so obvious as the above. We shall not go into the details of these cases, but shall merely list some of them below with appropriate references:

(a)   The case of linear taper $h = h_o(1 - r/a)$, $a > r_{outer}$ admits of a simple solution[10] when $\mu = 1/3$.

(b)   A power series solution[11] is possible for an annular plate with the taper defined by $h = h_o e^{-\beta r^2}$, which can be used as a good approximation for any smooth monotonically increasing or decreasing thickness by choosing an appropriate value for β.

(c)   While the general solution for the equation of motion of a uniform circular or annular plate is in terms of Bessel functions and hence frequency analysis requires the solution for the roots of a transcendental equation involving Bessel functions, it has been shown that much simpler solutions are possible for some variable thickness plates yielding the frequencies or mode shapes in closed form even for modes that are not axisymmetric. These cases include a completely

---

[10]R. Szilard, Theories and Applications of Plate Analysis, John Wiley and Sons, 2004.

[11]See, S. P. Timoshenko, S. W. Krieger, Theory of Plates and Shells, McGraw-Hill, 1959.

free circular plate[12] with $h = h_0[1 - (r/r_{\text{outer}})^2]$ and an isotropic or polar orthotropic annular plate with a parabolic taper $h = h_o r^2$ and with arbitrary edge conditions at the inner and outer boundaries.[13]

Inspired by the above solutions, and following an innovative inverse method wherein the mode shape is assumed a priori and the corresponding stiffness variation is found out, a series of exact closed-form solutions have been obtained recently by Elishakoff and his colleagues for both vibration and buckling problems.[14]

Summary

After pointing out the difficulties with respect to the analysis of variable thickness plates, some interesting exact solutions have been briefly discussed. It should be noted that exact solutions are possible for very few cases and are useful mainly for benchmarking various approximate methods; for practical problems involving variable thickness, there is no alternative except to employ some well-proven approximate method.

---

[12]G. Z. Harris, The normal modes of a circular plate of variable thickness, Quarterly Jl. of Mechanics and Applied Mathematics, 21, 1968, 32–36.

[13]T. A. Lenox, H. D. Conway, An exact, closed form, solution for the flexural vibration of a thin annular plate having a parabolic thickness variation, Jl. of Sound and Vibration, 68, 1980, 231–239.

[14]See, for example:

J. A. Storch, I. Elishakoff, Apparently first closed-form solutions of inhomogeneous circular plates in 200 years after Chladni, Jl. of Sound and Vibration, 276, 2004, 1108–1114;

I. Elishakoff, G. C. Ruta, Y. Stavsky, A novel formulation leading to closed-form solutions for buckling of circular plates, Acta Mechanica, 185, 2006, 81–88.

Chapter 13
Plate Buckling Due to Non-uniform Compression

The discussion of linear stability analysis in Chap.7 was confined to plates subjected to uniform edge loading. However, in practice, one encounters several cases where loads are not uniformly distributed along the edges. For such problems, stability analysis is more complicated because it requires the solution for the in-plane force field as a preliminary step. It is the purpose of the present chapter to illustrate this two-step procedure with specific reference to uniaxially compressed rectangular plates. The chapter includes converged results based on analytical solutions and presented in tabular form for future comparisons.[1]

13.1 The In-plane Problem

As pointed out in Chap. 7, the internal in-plane forces simply take their values at the boundary when a rectangular plate is subjected to uniform edge loads. For non-uniform edge loading, however, a formal solution for the in-plane force field is required from the following equilibrium equations derived in Sect. 7.1.

$$N_{x,x} + N_{xy,y} = 0$$
$$N_{xy,x} + N_{y,y} = 0 \tag{13.1}$$

with N_{yx} taken equal to N_{xy} as per Eq. (7.3).

---

[1] These results, as well as the mathematical procedure itself, have been reproduced here with copyright permission from: Prasun Jana, K. Bhaskar, Stability analysis of simply supported rectangular plates under non-uniform uniaxial compression using rigorous and approximate plane-stress solutions, Thin-Walled Structures, 44, 2006, 507–516 © Elsevier Ltd.

© The Author(s) 2021
K. Bhaskar and T. K. Varadan, *Plates*,
https://doi.org/10.1007/978-3-030-69424-1_13

This corresponds to the plane elasticity problem usually posed in terms of the in-plane stresses σ_x, σ_y and τ_{xy}; if one seeks a solution for these stresses using a stress approach, then an additional compatibility equation also needs to be considered to ensure a single-valued displacement field and thus preservation of the continuum nature of the domain after deformation. This was explained earlier in Sect. 10.1.1 with reference to the plane strain problem of the cantilevered plate strip, and it was shown that the problem could be reduced to the solution of a biharmonic equation in terms of Airy's stress function.

For the present case of plane stress without body forces, one can express the strain compatibility equation in terms of stresses following a procedure similar to that used in Sect. 10.1.1 and obtain, once again,

$$\nabla^2(\sigma_x + \sigma_y) = 0 \tag{13.2}$$

Employing the Airy's stress function ϕ, now defined by

$$N_x = h\sigma_x = h\phi_{,yy}, \, N_y = h\sigma_y = h\phi_{,xx}, \, N_{xy} = h\tau_{xy} = -h\phi_{,xy} \tag{13.3}$$

one can satisfy the equilibrium equations (Eq. (13.1)) identically and reduce the problem once again to the biharmonic equation

$$\nabla^4\varphi = 0 \tag{13.4}$$

to be solved with appropriate edge conditions.[2]

When a rectangular plate is subjected to normal and shear loads uniformly distributed along the edges, the solution is simply an identical, uniform force field throughout the domain, and this is clearly seen to satisfy the biharmonic equation and the edge traction conditions. A similar situation prevails when the applied loads vary linearly along the edges—for example, for a plate subjected to in-plane bending as shown in Fig. 13.1. However, when the tractions vary non-linearly or are applied on discrete portions of the boundary, the solution is not trivial.

For plates subjected to non-uniform uniaxial compression, we shall present a method based on superposition of three Fourier series solutions for ϕ, noting that such a multiple Fourier series method is one of the classical approaches suggested for plane elasticity problems.[3]

Taking the plate dimensions as $a \times b$, and the origin at the centre, we shall consider uniaxial compression due to normal stresses applied on the edges $x = \pm a/2$ and varying symmetrically with respect to the x-axis (Fig. 13.2). Without loss of generality, these applied stresses can be expressed in a Fourier cosine series as

---

[2]Thus, in the absence of body forces, the governing equation in terms of is the same for both plane stress and plane strain problems; the implications of this may be found in any standard book on the theory of elasticity.

[3]R. W. Little, Elasticity, Prentice-Hall, 1973.

Fig. 13.1 In-plane bending

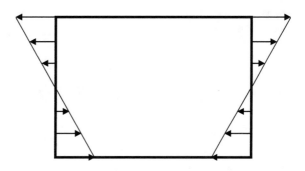

Fig. 13.2 Present problem

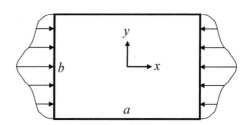

$$\sigma_x = \sum_{r=1,3,5..} \sigma_r \cos \frac{r\pi y}{b} \quad \text{at } x = \pm a/2 \qquad (13.5)$$

with sufficient terms taken to accurately represent the actual distribution.

A solution for ϕ_1, the first building block corresponding to the above σ_x, can be sought as

$$\phi_1 = \sum_{r=1,3,5} f_r(x) \cos \frac{r\pi y}{b} \qquad (13.6)$$

By substituting this in the biharmonic equation, one gets the ordinary differential equation

$$f_r^{iv} - 2\left(\frac{r\pi}{b}\right)^2 f_r'' + \left(\frac{r\pi}{b}\right)^4 = 0$$

The solution for this equation is easily obtained as

$$f_r = C_1, \cosh \frac{r\pi x}{b} + C_{2r} \sinh \frac{r\pi x}{b} + C_3, x \cosh \frac{r\pi x}{b} + C_4, x \sinh \frac{r\pi x}{b} \qquad (13.7)$$

Enforcement of the normal stress condition (Eq. 13.5) along with zero shear conditions at the edges $x = \pm a/2$ yields the constants as

Fig. 13.3 First building
block—edge stresses

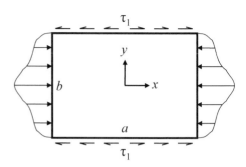

$$C_{1r} = \frac{\sigma_r b^2 \left(ar\pi \cosh \frac{ar\pi}{2b} + 2b \sinh \frac{ar\pi}{2b}\right)}{r^2\pi^2\left(ar\pi + b \sinh \frac{ar\pi}{b}\right)},$$

$$C_{4r} = -\frac{2\sigma_r b^2 \sinh \frac{ar\pi}{2b}}{r\pi\left(ar\pi + b \sinh \frac{ar\pi}{b}\right)}, \quad C_{2r} = C_{3r} = 0 \qquad (13.8)$$

Corresponding to this first building block, the normal stresses at the edges
$y = \pm b/2$ are identically zero while there are non-zero shear stresses as given
by

$$\tau_1 = \pm \sum_{r=1,3,5} \frac{r\pi}{b} \sin \frac{r\pi}{2} \left[\left(C_{1r}\frac{r\pi}{b} + C_{4r}\right) \sinh \frac{r\pi x}{b} + C_{4r}\frac{r\pi}{b}x \cosh \frac{r\pi x}{b}\right]$$

$$(13.9)$$

and as shown in Fig. 13.3. These stresses are distributed antisymmetrically with
respect to the y-axis and need to be eliminated.

For this purpose, the second building block is chosen to correspond to applied
shear stresses expressed as

$$\tau_{xy2} = \pm \sum_{n=1,3,5..} A_n \sin \frac{n\pi x}{a} \quad \text{at } y = \pm b/2, \text{ respectively} \qquad (13.10)$$

and zero normal stresses at these edges. Noting that the stress function corresponding
to the applied shear stresses should have a cosine variation in the x-direction and
proceeding exactly as was done for the first building block, one can reduce the
biharmonic equation to an ordinary differential equation that is readily solved. The
stress function ϕ_2 so obtained is

$$\phi_2 = \sum_{n=1,3,5} \cos \frac{n\pi x}{a} \left(C_{1n} \cosh \frac{n\pi y}{a} + C_{4n} y \sinh \frac{n\pi y}{a}\right) \qquad (13.11)$$

with

Fig. 13.4 Second building block—edge stresses

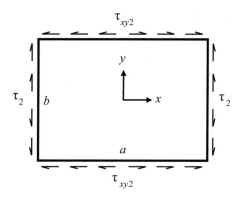

$$C_{1n} = -\frac{a^2 b A_n \sinh \frac{n\pi b}{2a}}{bn^2\pi^2 + an\pi \sinh \frac{bn\pi}{a}},$$

$$C_{4n} = \frac{2a^2 A_n \cosh \frac{n\pi b}{2a}}{bn^2\pi^2 + an\pi \sinh \frac{bn\pi}{a}} \qquad (13.12)$$

This building block produces no normal stresses at $x = \pm a/2$, but non-zero shear stresses as given by

$$\tau_2 = \pm \sum_{n=1,3,5} \frac{n\pi}{a} \sin \frac{n\pi}{2}\left[\left(C_{1n}\frac{n\pi}{a} + C_{4n}\right)\sinh \frac{n\pi y}{a} + C_{4n}\frac{n\pi}{a} y \cosh \frac{n\pi y}{a}\right]$$

$$(13.13)$$

and as shown in Fig. 13.4; these need to be eliminated by yet another building block. Noting that τ_2 is antisymmetric about the x-axis, ϕ_3 is taken to correspond to

$$\tau_{xy3} = \pm \sum_{m=1,3,5..} B_m \sin \frac{m\pi y}{b} \quad \text{at } x = \pm a/2 \qquad (13.14)$$

Using the biharmonic equation as done earlier, ϕ_3 is obtained as

$$\phi_3 = \sum_{m=1,3,5} \cos \frac{m\pi y}{b}\left(C_{1m}\cosh \frac{m\pi x}{b} + C_{4m}x \sinh \frac{m\pi x}{b}\right) \qquad (13.15)$$

with

$$C_{1m} = -\frac{ab^2 B_m \sinh(\frac{m\pi a}{2b})}{am^2\pi^2 + bm\pi \sinh(\frac{am\pi}{b})},$$

$$C_{4m} = \frac{2b^2 B_m \cosh(\frac{m\pi a}{2b})}{am^2\pi^2 + bm\pi \sinh(\frac{am\pi}{b})} \qquad (13.16)$$

Fig. 13.5 Third building
block—edge stresses

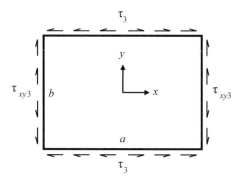

Fig. 13.5 Third building
block—edge stresses

so as to satisfy the zero normal stress condition along with the conditions of
Eq. (13.14) at $x = \pm a/2$.

This building block produces a non-zero shear stress at $y = \pm b/2$, as given by

$$\tau_3 = \pm \sum_{m=1,3,5} \frac{m\pi}{b} \sin \frac{m\pi}{2} \left[\left(C_{1m} \frac{m\pi}{b} + C_{4m} \right) \sinh \frac{m\pi x}{b} + C_{4m} \frac{m\pi}{b} x \cosh \frac{m\pi x}{b} \right]$$

(13.17)

and as shown in Fig. 13.5.

By superposing the three building blocks and applying zero shear boundary
conditions at the edges, one gets

$$\tau_1 + \tau_3 + \tau_{xy2} = 0 \text{ at } y = \pm b/2$$
$$\tau_2 + \tau_{xy3} = 0 \text{ at } x = \pm a/2 \tag{13.18}$$

With τ_1 and τ_3 expanded in a sine series of the form used in Eq. (13.10) and τ_2
as in Eq. (13.14), the above equations yield as many algebraic equations as the total
number of constants A_n and B_m used and thus these constants can be determined.

For illustrative purposes, four uniaxial distributions are considered here, as shown
in Fig. 13.6. It should be pointed out that load cases C, D have to be looked upon
as the superposition of a uniform load minus the centrally applied loads of cases
A, B so that the series representation of Eq. (13.5) is applicable. The number of
unknown constants A_n and B_m can be decided based on convergence of the shear
stress at any point of the edge to as small a value as desired. Taking the number of
A_n's to be equal to the number of B_m's for convenience [i.e. n_{max} in Eq. (13.10) =
m_{max} in Eq. (13.14)], and considering square plates, it can be shown that the edge
shear stress values rapidly decay with increasing m_{max}; for example, for a square
plate subjected to triangular load with a maximum intensity of σ_o as in Fig. 13.6
(Case B), the maximum shear stress at the edge decays to a value of the order of
$\sigma_o/1000$ for $m_{max} = 7$ and to that of the order of $\sigma_o/10,000$ for $m_{max} = 15$. It can
also be shown that the corresponding values of the in-plane stresses at any interior

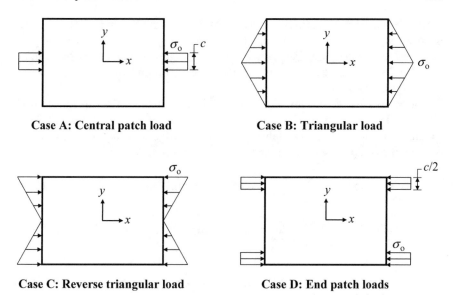

Fig. 13.6 Different load cases considered

point exhibit four-digit convergence. Thus, the final solution by this superposition approach, which satisfies the governing equation (Eq. 13.4) exactly, all the normal stress conditions on the edges exactly and the zero shear conditions on the edges to any desired degree of numerical accuracy, can be termed an exact solution.

By St. Venant's principle, any edge load distribution symmetric about the x-axis would diffuse into a uniform distribution in the interior of the plate far away from the loaded ends. However, such complete diffusion is possible only for a long plate ($a/b > 2$). For shorter plates, the stress field is characterized by a non-uniform distribution of σ_x everywhere including the interior and also by the presence of the other stress components σ_y and τ_{xy} of significant magnitudes; for example, for a square plate with triangular loading Fig. 13.6, it can be shown that these other components can be as high as $\sigma_o/2$.

13.2 Determination of the Critical Load

The governing equation for the buckled plate is

$$D\nabla^4 w = N_x w_{,xx} + N_y w_{,yy} + 2N_{xy} w_{,xy} \tag{13.19}$$

where the in-plane forces are now functions of x and y. Such a partial differential equation with variable coefficients does not admit of an exact solution in general. (The only case amenable to rigorous analysis is that of a plate with two simply supported opposite edges, say the edges parallel to the y-axis, subjected to linearly varying N_x over their entire length; using Levy's method, this problem can be reduced to an ordinary differential equation with variable coefficients and solved using a power series approach.[4])

For the present case with N_x, N_y and N_{xy} all in terms of a superposition of three Fourier series, an approximate solution of the buckling problem is called for. Confining attention to plates simply supported on all edges and employing Galerkin's method, the complete sets of trigonometric comparison functions

$$w(x, y) = \sum_{m=1,3,5..} \sum_{n=1,3,5..} W_{mn} \cos(m\pi x/a) \cos(n\pi y/b) \qquad (13.20)$$

$$w(x, y) = \sum_{m=1,2,3..} \sum_{n=1,3,5..} W_{mn} \sin(2m\pi x/a) \cos(n\pi y/b) \qquad (13.21)$$

can be used for modes symmetric and antisymmetric, respectively, with respect to the y-axis, and their choice is based on the expected buckled shape. As shown in Chap. 7 for uniform compression, the buckled shape depends on the aspect ratio of the plate; in addition, as will be shown here, it will also be influenced by the actual distribution of the edge load and hence it is often necessary to try out both the above sets of functions to identify the lowest buckling load. It should be pointed out that the lowest buckling load is expected to correspond to a buckled shape that is symmetric about the x-axis, but not a simple half-sine wave except for uniform compressive load, and hence trial functions with different odd harmonic wave numbers in the y-direction (i.e. values of n) are required.

As explained earlier in Sect. 8.3, the governing equation is satisfied in a weighted residual sense in Galerkin's method with the weighting functions taken to be the same as the approximating functions. This leads to an eigenvalue problem having non-trivial solutions only for certain discrete values of the applied load parameter which can be taken as σ_o (see Fig. 13.6); the lowest of these corresponds to the critical load. Using this value of σ_o, the relations between the different W_{mn} can be found out so that the buckled shape can be expressed using Eqs. (13.20) or (13.21). Since the chosen approximating functions form a complete set, convergence to the exact solution is guaranteed as the number of terms is increased.

Such converged results for the total compressive load are generated for square and rectangular plates with $a/b = 2$ with the chosen load distributions of Fig. 13.6. For the patch loads, two patch sizes are considered: $c = b/200$ to simulate a concentrated load, and $c = b/10$.

[4]J. H. Kang, A. W. Leissa, Exact solutions for the buckling of rectangular plates having linearly varying in-plane loading on two opposite simply supported edges, International Jl. of Solids and Structures, 42, 2005, 4220–4238.

Table. 13.1 Total buckling loads (values for K in $P_{cr} = K\pi^2 D/b$)

Load case	$a/b = 1.0$	$a/b = 2.0$
Uniform load	4.000	4.000
A1. Central concentrated load ($c = b/200$)	2.604	2.947
A2. Central patch load ($c = b/10$)	2.621	2.962
B. Triangular load	3.339	3.576
C. Reverse triangular load	4.934	4.350
D1. End patch load ($c = b/10$)	6.499	4.520
D2. End concentrated load ($c = b/200$)	6.541	4.580

Since the plane-stress solution is exact and the convergence of the Galerkin's method is guaranteed, the results obtained here will be useful for verifying the accuracy of various approximate numerical approaches. With this in mind, convergent results up to four significant digits are presented in tabular form (Table 13.1), with the values for uniform compression also included for the sake of completeness.

Focusing attention on the square plate first, one can see that the buckling load varies drastically for different load cases, from a minimum for a central concentrated load to a maximum for the plate with two end-point loads. It should be noted that the load cases of Fig. 13.6 are listed in an order such that the compressive action of the load moves gradually from the centre of the edge to the corners, and, as expected, the corresponding total buckling load increases in the same order. The actual edge load distribution has a significant influence on the critical value of the total load; even the slight change from uniform distribution to a triangular distribution changes the value by as much as 17%. In all these cases, the lowest buckling load corresponds to a shape with no nodal lines parallel to the y-axis.

For the longer plate with $a/b = 2$, it is expected that the difference between various load cases would be smaller than that for the square plate because of greater diffusion towards a uniform state of stress, and the results show that this is indeed true. For all the cases here, the lowest buckling load corresponds to a shape with one nodal line parallel to the y-axis, except for Case D2, for which there are no nodal lines.

Thus, the number of nodal lines parallel to the y-axis depends not only on the a/b ratio but also on the edge load distribution. It is well known that the uniformly compressed plate buckles into nearly square panels, with the transition points, which correspond to two equally probable buckled configurations differing by one nodal line for the same buckling load, occurring at $a/b = \sqrt{2}$, $\sqrt{6}$, etc. (Fig. 7.3). When the load is non-uniform, there is a shift towards left or right of the transition points, depending on whether the load intensity is greater or smaller, respectively, at the centre of the loaded edge.

13.3   Some Other Approaches

The method described above is based on a rigorous series solution for the in-plane problem and a convergent Galerkin's solution for the buckling problem. It is possible to carry either of these two steps or both using other approximate or numerical methods such as the finite element method, and several such solutions can be found in the literature. There have also been attempts to reduce the problem to a single-step procedure, and two such approaches are briefly mentioned below.

Let us confine attention to the example of a rectangular plate ($0 \leq x \leq a, 0 \leq y \leq b$) compressed by central point loads P applied on the edges $x = 0$ and $x = a$. For this problem, the method of Timoshenko[5] is based on equating the work done by the two forces as they approach each other to the strain energy of bending. This amounts to the use of the following functional with Rayleigh–Ritz method.

$$
\Pi = \frac{D}{2} \int [(\nabla^2 w)^2 - 2(1 - \mu)(w_{,xx} w_{,yy} - w_{,xy}^2)]\, \mathrm{d}x\, \mathrm{d}y - P \int_{x=0}^{x=a} \left. \frac{w_{,x}^2}{2} \right|_{y=\frac{b}{2}} \mathrm{d}x
\tag{13.22}
$$

By comparing this with the correct functional for the stability problem as given in Eq. (8.4), one can see that Timoshenko's method amounts to the assumption that the internal stress field coincides with the applied load yielding a uniform N_x field throughout the domain and zero N_y and N_{xy}.[6] It has been shown[7] that this method leads to a severe under-prediction of the total critical load. Further, for the case of end concentrated loads (load case D2 of Table 13.1), this method predicts the critical load to be infinitely large, a result totally unrealistic.

A more complicated method, put forth by Alfutov and Balabukh,[8] is based on a modified functional involving, besides the strain energy of bending, an integral depending on an assumed stress field which satisfies equilibrium equations and the applied in-plane load conditions but violating compatibility, and another integral depending on the above stress field as well as a stress function which satisfies the compatibility equation of von Karman's moderately large deformation theory (see Eq. (14.38)). Corresponding to an assumed one-term approximation of the buckled shape, it has been shown[9] that such a stress function can be easily determined and

---

[5]S. P. Timoshenko, J. M. Gere, Theory of Elastic Stability, McGraw-Hill, 1963.

[6]For other possible interpretations of this functional, see: N. A. Alfutov, Stability of Elastic Structures, Springer, 2000.

[7]A. W. Leissa, E. F. Ayoub, Vibration and buckling of a simply supported rectangular plate subjected to a pair of in-plane concentrated forces, Jl. of Sound and Vibration, 127, 1988, 155–171.

[8]N. A. Alfutov, L. I. Balabukh, On the possibility of solving plate stability problems without a preliminary determination of the initial state of stress, Jl. of Applied Mathematics and Mechanics (PMM), 31(4), 1967, 730–736.

[9]H. H. Spencer, H. Surjanhata, The simplified buckling criterion applied to plates with partial edge loading, Applied Scientific Research, 43, 1986, 79–90.

this leads to a very simple formula for the critical load; the resulting values have been shown to be quite accurate for the problem of central patch loading with any patch size ranging from 0 (point load) to b (uniform load).

Summary

With reference to uniaxial compression, a method for accurate plane-stress analysis and the use of the resulting in-plane force field for buckling analysis have been illustrated in this chapter. The dependence of the critical value on the actual edge load variation has been highlighted. The approach used herein can be directly extended to the case of biaxial loading[10] or material orthotropy.[11]

---

[10]Prasun Jana, K.Bhaskar, Analytical solutions for buckling of rectangular plates under non-uniform biaxial compression or uniaxial compression with in-plane lateral restraint, International Jl. of Mechanical Sciences, 49, 2007, 1104–1112.

[11]J. Bharat Kalyan, K. Bhaskar, An analytical parametric study on buckling of non-uniformly compressed orthotropic rectangular plates, Composite Structures, 82, 2008, 10–18.

Chapter 14
Non-linear Flexure and Vibrations

Our focus in this chapter is on problems wherein non-linearity arises solely due to large deformations, often referred to as *geometric non-linearity*. The other type of non-linearity arising due to a non-linear stress–strain relationship is called *material non-linearity,* and this will not be considered here. Geometric non-linearity itself can arise in two ways, and these are best understood with reference to the simple case of cylindrical bending which will be discussed first. The analysis is later extended to finite rectangular plates, and salient features associated with such non-linear behaviour are highlighted.

14.1 Cylindrical Bending of a Simply Supported Plate Strip

14.1.1 Case (a): Immovable Edges

Consider an infinitely long thin plate strip of thickness h with the longitudinal edges simply supported such that they do not move towards each other as the plate undergoes cylindrical bending (Fig. 4.1). Thus, bending would be accompanied by mid-plane stretching that becomes severe as the plate deflections increase. Let us analyse such a problem when the plate is subjected to a central line load of intensity P (i.e. force per unit length in the y-direction) as shown.

To capture the stiffening effect of mid-plane stretching, one has to consider the deformed configuration of the plate; with reference to this, the bending moment at any section can be written as

© The Author(s) 2021
K. Bhaskar and T. K. Varadan, *Plates*,
https://doi.org/10.1007/978-3-030-69424-1_14

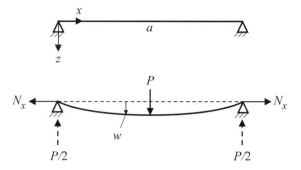

$$M_x(x) = \frac{P}{2}x - N_x w \quad \text{for } 0 \le x \le a/2 \tag{14.1}$$

where N_x is the membrane force intensity (i.e. force per unit length in the y-direction). The moment–curvature relation for this cylindrical bending problem leads to

$$-Dw_{,xx} + N_x w = \frac{P}{2}x$$

$$\text{i.e. } w_{,xx} - \frac{N_x}{D}w = -\frac{Px}{2D} \quad \text{for } 0 \le x \le a/2 \tag{14.2}$$

along with the two relevant boundary conditions

$$w(0) = 0$$

$$w_{,x}(a/2) = 0.$$

The solution for the above equation is

$$w = -\frac{Pa^3}{16\lambda^3 D}\text{sech}\lambda \ \sinh\frac{2\lambda x}{a} + \frac{Pa^2 x}{8\lambda^2 D} \quad \text{for } 0 \le x \le a/2 \tag{14.3}$$

where $\lambda = \sqrt{\frac{N_x a^2}{4D}}$ is the membrane force parameter.

The membrane force N_x corresponds to the change in length of the mid-plane between the two supports as it deforms. To find this, let us consider a small element of length dx moving from its undeformed position AB to the final position $A_1 B_1$ involving x-displacements u and $(u + u_{,x}\,dx)$, respectively, and z-displacements w and $(w + w_{,x}\,dx)$, respectively, for points A and B (Fig. 14.2).

Thus, the effect of the difference in x-displacements is to produce an elongation equal to $u_{,x}\,dx$, while the effect of the difference in w-displacements is to produce an elongation given approximately by

Fig. 14.2 Kinematics of deformation

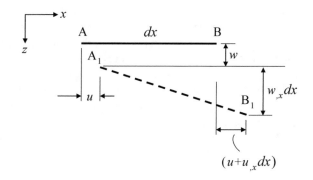

$$\sqrt{dx^2 + (w_{,x}dx)^2} - dx \approx \frac{w_{,x}^2}{2}dx. \tag{14.4}$$

Thus, the total elongation of the plate of original length a is given by

$$\int_0^a (u_{,x} + w_{,x}^2/2)\,dx = u\big|_0^a + \int_0^a (w_{,x}^2/2)\,dx = 2\int_0^{a/2} (w_{,x}^2/2)\,dx \tag{14.5}$$

since the ends are immovable and w is symmetric about mid-span.

Equating this to the change in length as arising due to a uniform N_x along the span, one gets

$$2\int_0^{a/2} (w_{,x}^2/2)\,dx = \frac{(N_x/h)a}{\{E/(1-\mu^2)\}} \tag{14.6}$$

where the right-hand-side expression follows from the plane-stress constitutive law corresponding to $\sigma_z = 0$ [Eq. (2.9a)], along with $\varepsilon_y = 0$.

Substituting for w from Eq. (14.3) and rewriting N_x in terms of λ, one gets, after rearrangement,

$$\left(\frac{Pa^3}{Dh}\right)^2 = \frac{256\lambda^7}{3[\lambda(2 + \mathrm{sech}^2\lambda) - 3\tanh\lambda]} \tag{14.7}$$

which relates λ, the non-dimensional membrane force parameter, to the non-dimensional load parameter (Pa^3/Dh). For any load, this equation has to be solved iteratively for λ, and then the deflection is obtained using Eq. (14.3). The maximum stress in the plate, occurring at the bottommost point at mid-span, is the sum of the membrane component $\frac{N_x}{h}$ and the bending component $\frac{6}{h^2}\left(\frac{Pa}{4} - N_x w_{max}\right)$.

A look at Eqs. (14.3) and (14.7) clearly shows that the deflection is related non-linearly to the transverse load P, but because of the complicated expressions therein, one can get an idea of the nature and degree of non-linearity only by a numerical solution for different loads. Before we present such numerical results, let us solve this problem approximately using Rayleigh–Ritz method so as to obtain a simpler equation that directly reveals the nature of the non-linearity.

Assuming that the deflected shape can be approximated as a sine curve

$$w = A \sin \frac{\pi x}{a} \tag{14.8}$$

N_x can be calculated using Eq. (14.6) as

$$N_x = \frac{\pi^2 E h A^2}{4(1 - \mu^2)a^2} = \frac{3\pi^2 D}{a^2}\left(\frac{A}{h}\right)^2 \tag{14.9}$$

The total strain energy (per unit length of the infinite strip) is the sum of that due to stretching and that due to bending, as given by

$$U_{stretching} = \frac{1}{2}\left(\frac{N_x}{h}\right)\left(\frac{N_x}{h}\cdot\frac{1-\mu^2}{E}\right)ah$$

$$U_{bending} = \frac{D}{2}\int_0^a w_{,xx}^2\,dx \tag{14.10}$$

where the coefficient $\frac{1}{2}$ appears because the stresses and strains are linearly related. The corresponding potential energy of the applied load is given by

$$V = -P w|_{a/2} \tag{14.11}$$

Using Eqs. (14.8) and (14.9), the total potential energy can then be expressed in terms of A and is minimized by taking

$$\frac{d(U_{total} + V)}{dA} = 0 \tag{14.12}$$

to finally yield

$$\left(\frac{A}{h}\right)\left[1 + 3\left(\frac{A}{h}\right)^2\right] = \frac{2}{\pi^4}\left(\frac{Pa^3}{Dh}\right) \tag{14.13}$$

This equation clearly shows the hardening nature of the non-linearity; the load increment required to cause a unit increment in the maximum deflection is not constant but increases with the maximum deflection, i.e. the stiffness relating the central load P and the maximum deflection A increases as the deflections increase. The equation also shows that the effect of this non-linearity is small as long as the maximum deflection is small compared to the thickness of the plate.

A note about the accuracy of this approximate Rayleigh–Ritz solution is appropriate here. This accuracy is easily verified with respect to the linear problem corresponding to very small deflections and hence negligible N_x; in this case, the exact solution for the maximum deflection is $\frac{Pa^3}{48D} = 0.02083\frac{Pa^3}{D}$ while the approximate solution is $\frac{2Pa^3}{\pi^4 D} = 0.02053\frac{Pa^3}{D}$, an under-prediction by about 1.5%.

Table 14.1 presents numerical results for the non-linear problem based on the exact solution, with the load parameter taken as $(Pa^3/48Dh)$; it should be noted that this can also be written as $\frac{w_{max\ linear}}{h}$, and thus any value of this load parameter is the load required to cause $\frac{w_{max}}{h}$ of that magnitude as per the linear theory. The non-linear equation Eq. (14.7) is easily solved using *MATLAB* or *Mathematica*. The results are presented in the form of λ^2, $\frac{w_{max\ linear}}{w_{max}}$ and $\frac{\sigma_{max\ linear}}{\sigma_{max}}$, where $\sigma_{max\ linear}$ is $\frac{6}{h^2} \cdot \frac{Pa}{4}$. To show the increasing significance of the membrane action as the deflections increase, the ratio of the membrane component of σ_{max} to the bending component is also presented in Table 14.1.

The results based on Rayleigh–Ritz method are presented in Table 14.2; for the values of the load parameter taken as in Table 14.1, Eq. (14.13) is solved for (A/h) using *Mathematica*. For finding out the ratio of $w_{max\ linear}$ to w_{max}, the former is taken as that of the approximate solution, i.e. $\frac{2Pa^3}{\pi^4 D}$; as can be seen, this yields excellent correlation with the exact solution. This simply means that the percentage error

Table 14.1 Comparison of linear and non-linear analysis (exact solution)

$\frac{Pa^3}{48Dh} = \frac{w_{max\ linear}}{h}$	$\lambda^2 = \frac{N_x a^2}{4D}$	$\frac{w_{max\ linear}}{w_{max}}$	$\frac{\sigma_{max\ linear}}{\sigma_{max}}$	$\frac{\sigma_{membrane}}{\sigma_{bending}}$
0.1	0.0682	1.027	0.984	0.041
0.2	0.2394	1.096	0.992	0.072
0.3	0.4604	1.184	1.046	0.098
0.4	0.6997	1.279	1.093	0.12
0.5	0.9432	1.376	1.141	0.14

Table 14.2 Comparison of linear and non-linear analysis (Rayleigh–Ritz method)

$\frac{Pa^3}{48Dh}$	$\lambda^2 = \frac{N_x a^2}{4D}$	$\frac{w_{max\ linear}}{w_{max}}$
0.1	0.0681	1.028
0.2	0.2390	1.097
0.3	0.4598	1.186
0.4	0.6987	1.283
0.5	0.9416	1.382

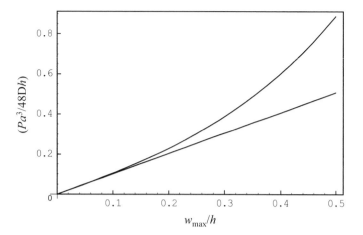

Fig. 14.3 Linear and non-linear load–deflection plots

of Rayleigh–Ritz method is almost the same for linear and non-linear deflections in this particular problem, and thus it gets neutralized when the deflection ratio is considered. Table 14.2 also shows that the membrane force is accurately predicted.

From the above analysis, it is clear that the load versus deflection behaviour in this problem is one of the hardening types, i.e. greater load increments are required for the same deflection increments as the deflections become larger.

The linear theory is found to over-predict the deflections by a few per cent for $w_{\text{max linear}}$ less than 10% of the plate thickness, and thereafter this error rapidly increases due to increasing influence of stretching. The linear and non-linear load–deflection curves based on Eq. (14.13) are as shown in Fig. 14.3.

From Table 14.1, it is seen that the maximum stress is also over-predicted by the linear theory at large deflections, but the errors are less than those for deflections. Thus, when a plate is so thin that its deflections are comparable to the thickness, the linear theory predictions for both deflections and stresses are on the conservative side, leading to a safe, over-designed plate.

To understand the relevance of non-linear analysis in practice, let us consider some specific examples. As a first step, one can relate the maximum deflection and the maximum bending stress, based on linear theory for convenience, as

$$\frac{w_{\text{max linear}}}{h} = \frac{Pa^3}{48Dh} = \left(\frac{4h^2\sigma_{\text{max linear}}}{6a}\right)\left(\frac{a^3}{48Dh}\right)$$

i.e. $\quad \dfrac{w_{\text{max linear}}}{h} = \dfrac{(1 - \mu^2)\,\sigma_{\text{max linear}}}{6}\dfrac{1}{E}\left(\dfrac{a}{h}\right)^2$

Using this, one can easily examine the magnitude of deflections when the maximum stress reaches the yield point. Considering three materials—a typical low-carbon steel with $E = 200$ GPa, $\sigma_{\text{yield}} = 200$ MPa; an alloy steel with $E = 200$ GPa, $\sigma_{\text{yield}} = 600$ MPa; and an aluminium alloy with $E = 70$ GPa, $\sigma_{\text{yield}} = 300$ MPa, with

$\mu = 0.3$ for all the three, one gets, respectively,

$$\frac{w_{\text{max linear}}}{h} = (0.15, 0.45, 0.64) \times 10^{-3} \left(\frac{a}{h}\right)^2$$

corresponding to the onset of yielding. Thus, considering two thickness ratios of the plate, one gets

$$\frac{w_{\text{max linear}}}{h} = (1.5, 4.5, 6.4) \text{ for } a/h = 100$$

$$= (0.6, 1.8, 2.6) \text{ for } a/h = 20$$

which clearly shows that, in general, deflections well beyond the thickness are possible within the elastic limit, the magnitude of such deflections being more severe for thinner plates and for materials with higher yield point or lower Young's modulus. It is also clear that none of the above example cases can be analysed with acceptable accuracy using the linear theory.

14.1.2 Case (b): Freely Movable Edges

As an alternative to the above problem, let us now consider a plate strip supported on two roller supports that are free to move towards each other as the plate bends, so that no membrane force is generated (Fig. 14.4).

The source of non-linearity here is with respect to the moment–curvature relationship. Noting that the present problem of cylindrical bending is the plane strain counterpart of a beam and hence that Kirchhoff–Poisson hypothesis reduces to the assumption of plane cross sections, the relation between the bending moment M_x and the curvature $(1/R_x)$ at any x (see Fig. 14.5) can be obtained as

$$M_x = \frac{D}{R_x} \tag{14.14}$$

Fig. 14.4 Plate strip with freely movable edges

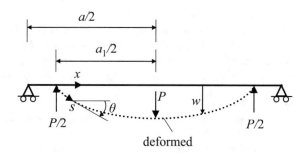

Fig. 14.5 Moment and
curvature

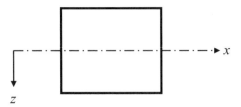

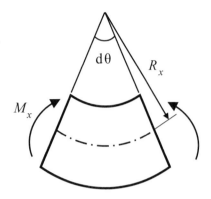

where the curvature, for arbitrarily large deflections, is given by

$$\frac{1}{R_x} = -\frac{w_{,xx}}{\left(1 + w_{,x}^2\right)^{\frac{3}{2}}}$$ (14.15)

instead of just $-w_{,xx}$ as used earlier (see Eq. [2.10)] in the linear theory.
The curvature can also be expressed in terms of the change in slope along the
deformed centroidal axis (see Figs. 14.4 and 14.5) as

$$\frac{1}{R_x} = -\frac{d\theta}{ds}$$ (14.16)

the negative sign indicating that θ decreases with s for positive curvature as shown.

Before proceeding ahead with the solution of the present simply supported strip
problem, let us note that it is equivalent to the solution of a cantilevered strip of half
the span with a tip load of $P/2$. An exact solution of this non-linear problem, for the
beam counterpart, was first obtained by Bisshopp and Drucker[1] in terms of elliptic
integrals. We shall adopt a slightly different approach, employed earlier by Varadan.[2]

---

[1]K.E. Bisshopp, D.C. Drucker, Large deflections of cantilever beams, Quarterly of Applied
Mathematics, 3, 1945, 272–275.

[2]T.K. Varadan, Non-linear bending of a cantilever beam of variable cross-section, Jl. of the
Aeronautical Society of India, 26, 1974, 1–5.

With origin for x chosen at the left end of the deformed configuration in Fig. 14.4, the bending moment at any section of the left half of the span is simply $Px/2$. Noting that $ds = \frac{dx}{\cos\theta}$, one has from Eqs. (14.14) and (14.16),

$$\cos\theta \, d\theta = -\frac{Px}{2D} dx \tag{14.17}$$

Integrating this between any x and mid-span, one gets

$$\sin\theta = \frac{P}{4D}\left(\frac{a_1^2}{4} - x^2\right) \tag{14.18}$$

where a_1 is the distance between the roller supports after deformation. The maximum slope occurring at the left end is given by

$$\sin\theta_{max} = \frac{Pa_1^2}{16D} \tag{14.19}$$

Corresponding to zero membrane force, the curved length of the centroidal axis should be equal to the undeformed length a. We have

$$ds = \frac{dx}{\cos\theta} = \frac{dx}{\sqrt{1 - \sin^2\theta}} = \frac{dx}{\sqrt{1 - \left[\frac{P}{4D}\left(\frac{a_1^2}{4} - x^2\right)\right]^2}}$$

which can be integrated between the left end and mid-span to yield

$$\frac{a}{2} = \int_0^{a_1/2} \frac{dx}{\sqrt{1 - \left[\frac{P}{4D}\left(\frac{a_1^2}{4} - x^2\right)\right]^2}} = \frac{a_1}{2}\int_0^1 \frac{d\xi}{\sqrt{1 - \eta^2\left(1 - \xi^2\right)^2}}$$

and hence,

$$\frac{a}{a_1} = \int_0^1 \frac{d\xi}{\sqrt{1 - \eta^2\left(1 - \xi^2\right)^2}} \tag{14.20}$$

where $\xi = x/(a_1/2)$, and η is the load parameter given by $(Pa_1^2/16D)$. To obtain the maximum deflection, we use

$$dw = \tan\theta \, dx = \frac{\sin\theta}{\sqrt{1 - \sin^2\theta}} dx = \frac{\eta(1 - \xi^2)}{\sqrt{1 - \eta^2(1 - \xi^2)^2}}\left(\frac{a_1}{2}\right)d\xi$$

and integrate this between the left end and mid-span to obtain

$$w_{max} = \frac{a_1}{2} \int_0^1 \frac{\eta(1 - \xi^2)}{\sqrt{1 - \eta^2(1 - \xi^2)^2}} d\xi \tag{14.21}$$

The maximum bending stress is given by

$$\sigma_x = \frac{6M_{max}}{h^2} = \frac{6}{h^2}\left(\frac{Pa_1}{4}\right) \tag{14.22}$$

which is less than the linear theory prediction based on $M_{max} = Pa/4$, i.e. the non-linear value is (a_1/a) times the linear estimate.

A look at Eqs. (14.19–14.21) reveals that they cannot be solved directly for the maximum slope, deflection, etc., in terms of load P; instead, for any value of the load parameter η, one has to first find a_1 using Eq. (14.20) and hence P, and thereafter the maximum slope, deflection and bending stress. Slopes and deflections at other points can be found in a similar manner. The integrals in Eqs. (14.20–14.21) can be evaluated numerically to any required accuracy and are obtained here using *Mathematica*.

Before presenting results based on the above rigorous analysis, let us once again employ Rayleigh–Ritz method to obtain an elegant equation revealing the nature of non-linearity.

The deflected shape is approximated as

$$w = A \sin \frac{\pi s}{a} \tag{14.23}$$

Noting that

$$\sin \theta = \frac{dw}{ds}$$

i.e.

$$\theta = \sin^{-1} w_{,s}$$

the curvature is given by

$$\frac{1}{R_x} = -\theta_{,s} = -\frac{w_{,ss}}{\sqrt{1 - w_{,s}^2}} \tag{14.24}$$

Thus, the strain energy per unit length of the infinite strip is given by

$$U_e = \frac{D}{2} \int \left(\frac{1}{R_x}\right)^2 ds = \frac{D}{2} \int_0^a \frac{w_{,ss}^2}{1 - w_{,s}^2} ds$$

$$= \frac{D}{2} \int_0^a w_{,ss}^2 (1 + w_{,s}^2 + w_{,s}^4 + \ldots) ds \qquad (14.25)$$

Use of this strain energy expression along with the approximation for w in Rayleigh–Ritz method yields

$$\left(\frac{A}{a}\right)\left[1 + \frac{\pi^2}{2}\left(\frac{A}{a}\right)^2 + \frac{3\pi^4}{8}\left(\frac{A}{a}\right)^4 + \ldots\right] = \frac{2}{\pi^4}\left(\frac{Pa^2}{D}\right)$$

i.e.

$$\left(\frac{A}{h}\right)\left[1 + \frac{\pi^2}{2}\left(\frac{A}{a}\right)^2\right] = \frac{2}{\pi^4}\left(\frac{Pa^3}{Dh}\right) \qquad (14.26)$$

assuming that higher powers of (A/a) can be neglected.

The above equation reveals the hardening nature of non-linearity clearly. It also shows that the non-linearity depends on the ratio of the maximum deflection to the span and is negligible as long as this ratio is small. This equation has to be contrasted with Eq. (14.13) applicable for the earlier case of immovable supports wherein departure from linearity begins as soon as the maximum deflection becomes comparable to the thickness of the plate. It should also be noted that both the equations turn out to be identical when the deflections are small and the non-linear terms become negligible, in which case one need not distinguish between immovable and freely movable edges.

We shall now verify the accuracy of Eq. (14.26) with respect to the exact solution based on Eq. (14.21). The results of the exact solution are first presented in Table 14.3; for various assumed values of the parameter η, after finding a_1 using Eq. (14.20), the load is expressed in terms of the parameter $(Pa^2/48D)$ which is equivalent to $(w_{\text{max linear}}/a)$. Later, the corresponding w_{max} is found out using Eq. (14.21).

For the Rayleigh–Ritz solution, the results for (A/a), corresponding to the values of the load used above, are obtained by solving Eq. (14.26) using *Mathematica* and are presented in the form of the ratio $\frac{w_{\text{max linear}}}{w_{\text{max}}}$ where the linear value, obtained by linearizing Eq. (14.26), is taken as $\frac{2Pa^3}{\pi^4 D}$. These approximate results are also presented in Table 14.3.

From this table, one can see that the maximum deflection predictions of the linear theory are acceptable as long as they are less than about 10% of the span a (this corresponds to $\eta = 0.3$); for larger deflections, linear theory errors increase rapidly. The maximum stress predictions by the linear theory are not as erroneous as those for

Table 14.3 Results for plate strip with freely movable edges

η	$\dfrac{a}{a_1}$	$\dfrac{Pa^2}{48D} = \dfrac{w_{max\ linear}}{a}$	$\dfrac{w_{max\ linear}}{w_{max}}$ (Exact)	$\dfrac{w_{max\ linear}}{w_{max}}$ (Rayleigh–Ritz)
0.1	1.003	0.0335	1.003	1.005
0.2	1.011	0.0681	1.018	1.021
0.3	1.025	0.105	1.044	1.048
0.4	1.047	0.146	1.082	1.087
0.5	1.078	0.194	1.138	1.139
0.6	1.123	0.252	1.220	1.209
0.7	1.188	0.329	1.344	1.305

Note: $\dfrac{a}{a_1} = \dfrac{\sigma_{max\ linear}}{\sigma_{max}}$

the maximum deflections, and are on the conservative side. Finally, the Rayleigh–Ritz solution is acceptable in that it leads to an adequate quantitative estimate of the non-linear effects, for the range of deflections considered here.

14.1.3 Observations from the Above Solutions

The above problems show that there are two sources of geometric non-linearity—one due to membrane forces that arise due to bending and the other due to the non-linear curvature term. Of these, the former occurs as soon as the plate deflections reach values comparable to the thickness while the latter occurs only when they are comparable to the in-plane dimensions. It should be recognized that the stretching effect is absent only for the case of cylindrical bending with freely movable supports, i.e. when the originally flat plate bends into a developable surface; for every other case, including that of a finite plate with freely movable edges, the bent configuration is non-developable and hence is associated with mid-plane stretching. The non-linear curvature effect is present in every problem but needs to be accounted for only when the deflections are quite large and comparable to the in-plane dimensions of the plate.

Thus, for a range of deflections usually termed *moderately large*, it is sufficient to account for the membrane forces alone with the curvature still approximately obtained by using the linear expression.

14.2 Moderately Large Deformation Theory

It is first appropriate to clearly distinguish between *small deformations* leading to the linear theory presented in Chap. 2 and *moderately large deformations*.

With reference to the analysis of plates, the term *small deformations* is used to imply that.

(a) All the strain components are negligibly small compared to unity. (They can be defined as fractional changes of length and deviations from original right angles only if this assumption is true.)

(b) Slopes are negligibly small compared to unity. (Thus, the approximate linear curvature expressions are valid, and, whenever necessary, sine and cosine of the slope at a point can be replaced by the slope itself in radians and unity, respectively.)

(c) Squares and products of slopes are negligible compared to the strain components. (This automatically implies that squares and products of $u_{,x}$, $u_{,y}$, $v_{,x}$ and $v_{,y}$ are also negligible and lead to linear strain–displacement relations as given by Eq. (1.2).)

These assumptions are true as long as the deflections of the plate are very small compared to its thickness.

By *moderately large deformations*, one implies assumptions (a) and (b) above, but not (c); instead, one assumes that.

(d) Squares and products of the in-plane displacement derivatives $u_{,x}$, $u_{,y}$, $v_{,x}$ and $v_{,y}$ are negligible compared to the in-plane strain components, and non-linear terms occurring in the expressions for the transverse strains are negligible.

Thus, while linear curvature expressions are retained here, the strain–displacement relations become non-linear, and this is the mathematical step required to capture the hardening effect due to the development of stretching forces during bending. Such a theory, often referred to after von Karman who put it forth in 1910, is applicable when the deflections range from a small fraction of the plate thickness to a few times its value, but remain very small as compared to the in-plane dimensions.

The starting point of this theory is the consideration of the strain–displacement relations corresponding to finite deformations of a three-dimensional body, derived from a study of kinematics alone,[3] as

$$\varepsilon_x = u_{,x} + \frac{1}{2}(u_{,x}^2 + v_{,x}^2 + w_{,x}^2)$$

$$\varepsilon_y = v_{,y} + \frac{1}{2}(u_{,y}^2 + v_{,y}^2 + w_{,y}^2)$$

$$\varepsilon_z = w_{,z} + \frac{1}{2}(u_{,z}^2 + v_{,z}^2 + w_{,z}^2) \tag{14.27a}$$

$$\gamma_{yz} = v_{,z} + w_{,y} + u_{,y}u_{,z} + v_{,y}v_{,z} + w_{,y}w_{,z}$$

$$\gamma_{xz} = u_{,z} + w_{,x} + u_{,x}u_{,z} + v_{,x}v_{,z} + w_{,x}w_{,z}$$

$$\gamma_{xy} = u_{,y} + v_{,x} + u_{,x}u_{,y} + v_{,x}v_{,y} + w_{,x}w_{,y} \tag{14.27b}$$

---

[3] See: A.R. Ragab, S.E. Bayoumi, Engineering Solid Mechanics, CRC Press, 1998.

often referred to as the components of Green's or Lagrangian strain tensor. By virtue of assumption (d) above, these equations are reduced to

$$\varepsilon_x = u_{,x} + \frac{1}{2}w_{,x}^2$$

$$\varepsilon_y = v_{,y} + \frac{1}{2}w_{,y}^2$$

$$\gamma_{xy} = u_{,y} + v_{,x} + w_{,x}w_{,y} \tag{14.28a}$$

$$\varepsilon_z = w_{,z}$$

$$\gamma_{yz} = v_{,z} + w_{,y}$$

$$\gamma_{xz} = u_{,z} + w_{,x} \tag{14.28b}$$

Equating γ_{xz}, γ_{yz} and ε_z to zero, as was done in Chap. 2, one obtains the displacement field. This procedure yields

$$u = u_o - zw_{,x}$$

$$v = v_o - zw_{,y}$$

$$w = w(x, y) \tag{14.29}$$

where $u_o(x,y)$ and $v_o(x,y)$ are introduced to account for mid-plane stretching.

The strain field is then obtained from Eq. (14.28a) as

$$\varepsilon_x = (u_{o,x} + \frac{1}{2}w_{,x}^2) - zw_{,xx} = \varepsilon_x^o - zw_{,xx}$$

$$\varepsilon_y = (v_{o,y} + \frac{1}{2}w_{,y}^2) - zw_{,yy} = \varepsilon_y^o - zw_{,yy}$$

$$\gamma_{xy} = (u_{o,y} + v_{o,x} + w_{,x}w_{,y}) - 2zw_{,xy} = \gamma_{xy}^o - 2zw_{,xy} \tag{14.30}$$

where ε_x^o, ε_y^o and γ_{xy}^o are the membrane components.

Using the plane-stress constitutive law, the stress field can be written as

$$\sigma_x = \frac{E}{(1 - \mu^2)}[(\varepsilon_x^o + \mu\varepsilon_y^o) - z(w_{,xx} + \mu w_{,yy})]$$

$$\sigma_y = \frac{E}{(1 - \mu^2)}[(\varepsilon_y^o + \mu\varepsilon_x^o) - z(w_{,yy} + \mu w_{,xx})]$$

$$\tau_{xy} = \frac{E}{2(1 + \mu)}(\gamma_{xy}^o - 2zw_{,xy}) \tag{14.31}$$

Comparing this with the stress field of the linear theory [Eq. (2.11)], one can see that there is an additional, z-symmetric, component in all the three stresses. Thus, besides the bending and twisting moments, in-plane normal and shear forces per unit length need to be included as stress resultants.

They are given by

$$(N_x, N_y, N_{xy}) = \int\limits_{-h/2}^{h/2} (\sigma_x, \sigma_y, \tau_{xy})\, dz \qquad (14.32)$$

It should be noted that N_{xy} acting on the edge normal to the x-axis is always accompanied by a complementary shear force $N_{yx} (=N_{xy})$ acting on the edge normal to the y-axis. The sign convention for these in-plane stress resultants follows directly from that for the three stresses (see Sect. 1.1).

The next step is to obtain the equilibrium equations in terms of the forces and moments by considering an elemental area of the deformed plate. This, however, has been done earlier in Sect. 7.1 while considering small deformations due to combined bending and stretching. The only difference between this earlier problem and the present one is that in the former, the internal forces N_x, N_y and N_{xy} depend only on the applied in-plane edge tractions and remain unchanged as the plate bends, while here they vary as the plate bends and are as yet undetermined. As has been pointed out [see assumption (b)], the approximations given by $\sin\theta \cong \theta$ and $\cos\theta \cong 1$, employed in Sect. 7.1 based on the small deformation assumption, are true for the present case of moderately large deformations also.

For the sake of completeness, the equilibrium equations are reproduced below.

$$N_{x,x} + N_{xy,y} = 0$$

$$N_{xy,x} + N_{y,y} = 0$$

$$Q_{x,x} + Q_{y,y} + q + N_x w_{,xx} + N_y w_{,yy} + 2N_{xy} w_{,xy} = 0$$

$$M_{xy,x} + M_{y,y} - Q_y = 0$$

$$M_{x,x} + M_{xy,y} - Q_x = 0 \qquad (14.33)$$

By eliminating Q_x and Q_y, the above equations can be reduced to

$$N_{x,x} + N_{xy,y} = 0$$

$$N_{xy,x} + N_{y,y} = 0$$

$$M_{x,xx} + 2M_{xy,xy} + M_{y,yy}$$
$$+ q + N_x w_{,xx} + N_y w_{,yy} + 2N_{xy} w_{,xy} = 0 \qquad (14.34)$$

The in-plane forces may be expressed in terms of displacements u_o, v_o and w using Eqs. (14.30–14.32) as

$$N_x = \frac{Eh}{(1-\mu^2)}[(u_{o,x} + \frac{1}{2}w_{,x}^2) + \mu(v_{o,y} + \frac{1}{2}w_{,y}^2)]$$

$$N_y = \frac{Eh}{(1-\mu^2)}[(v_{o,y} + \frac{1}{2}w_{,y}^2) + \mu(u_{o,x} + \frac{1}{2}w_{,x}^2)]$$

$$N_{xy} = \frac{Eh}{2(1+\mu)}(u_{o,y} + v_{o,x} + w_{,x}w_{,y}) \tag{14.35}$$

Using these and the moment–curvature relations [Eqs. (2.13)], the equilibrium equations may be expressed in terms of the displacements to obtain the governing equations of the moderately large deformation problem. As part of the displacement approach, such equations can be solved along with appropriate boundary conditions.

Alternatively, as is commonly done in the literature on large deformation problems, one may express the in-plane forces in terms of Airy's stress function ϕ as

$$N_x = h\phi_{,yy}$$
$$N_y = h\phi_{,xx}$$
$$N_{xy} = -h\phi_{,xy} \tag{14.36}$$

so that the first two equilibrium equations are identically satisfied.

The stress function should be such that it satisfies an equation corresponding to compatibility of the membrane strains. Noting that

$$\varepsilon_x^o = (u_{o,x} + \frac{1}{2}w_{,x}^2)$$

$$\varepsilon_y^o = (v_{o,y} + \frac{1}{2}w_{,y}^2)$$

$$\gamma_{xy}^o = (u_{o,y} + v_{o,x} + w_{,x}w_{,y}) \tag{14.37}$$

and eliminating u_o and v_o, one gets the strain compatibility equation as

$$\varepsilon_{x,yy}^o + \varepsilon_{y,xx}^o - \gamma_{xy,xy}^o = w_{,xy}^2 - w_{,xx}w_{,yy} \tag{14.38}$$

This can be expressed in terms of the membrane forces and hence the stress function using

$$\varepsilon_x^o = \frac{N_x - \mu N_y}{hE}, \quad \varepsilon_y^o = \frac{N_y - \mu N_x}{hE}, \quad \gamma_{xy}^o = \frac{2(1+\mu)N_{xy}}{hE} \tag{14.39}$$

to finally yield

$$\nabla^4\phi = Eh(w_{,xy}^2 - w_{,xx}w_{,yy}) \tag{14.40}$$

as an equation to be solved along with the third equilibrium equation [last of Eqs. (14.34)] which can also be expressed in terms of w and ϕ as

$$D\nabla^4 w = q + h\phi_{,yy}w_{,xx} + h\phi_{,xx}w_{,yy} - 2h\phi_{,xy}w_{,xy} \tag{14.41}$$

Equations (14.40) and (14.41) are referred to as von Karman equations and represent the two-way coupling between w and ϕ, i.e. deflections dependent on the membrane forces and membrane forces dependent on the deflections.

The boundary conditions of this theory involve both in-plane conditions and those for the bending problem. We shall not list all the possible combinations here, but shall merely point out that the classical in-plane edge conditions amount to either complete suppression of the tangential/normal displacement or complete absence of the corresponding shear/normal force, while several non-classical combinations are possible. This will be clear from some of the examples discussed below.

It is easy to show that the problem of cylindrical bending of an infinite strip with immovable edges considered earlier (Sect. 14.1.1) can be solved using the above von Karman equations with all y-derivative terms neglected, and this approach leads to the same solution as obtained earlier. The case of a finite rectangular plate is discussed below.

14.3   Flexure of a Simply Supported Rectangular Plate

An exact solution[4] is available for this problem wherein the governing equations are reduced to a doubly infinite system of non-linear algebraic equations and solved by appropriate truncation. This solution is too complicated for inclusion here; further, it does not provide an elegant final equation that reveals the degree of non-linearity. For this reason, we shall discuss a simpler approximate solution.

The problem considered is that of a rectangular plate ($0 \le x \le a, 0 \le y \le b$) with simply supported edges defined by

$$w = w_{,xx} = 0 \quad \text{at } x = 0, a$$

$$w = w_{,yy} = 0 \quad \text{at } y = 0, b \tag{14.42}$$

along with one of the following two sets of in-plane conditions

---

[4]S.Levy, Bending of rectangular plates with large deflections, NACA Report No.737, 1942 (*available online for open access*).

$$u = N_{xy} = 0 \quad \text{at } x = 0, a$$

$$v = N_{xy} = 0 \quad \text{at } y = 0, b \tag{14.43a}$$

$$u = \text{constant along } y, \text{ with } \int_{y=0}^{b} N_x dy = N_{xy} = 0 \quad \text{at } x = 0, a$$

$$v = \text{constant along } x, \text{ with } \int_{x=0}^{a} N_y dx = N_{xy} = 0 \quad \text{at } x = 0, b \tag{14.43b}$$

Of the two alternatives above, the former corresponds to immovable edges while the latter to freely movable straight edges. (A practical example involving the latter conditions is that of a rectangular portion of an orthogonally stiffened panel, bounded on its four sides by thin-walled open cross-sectional stiffeners having high bending rigidity but low torsional rigidity.) Both the immovable and movable edges considered above are free of in-plane shear forces, and hence the in-plane tangential displacement is freely permitted.

Considering uniform transverse load $q(x,y) = q_o$, and assuming the deflection to be

$$w = W \sin \frac{\pi x}{a} \sin \frac{\pi y}{b} \tag{14.44}$$

one gets from Eq. (14.40)

$$\nabla^4 \phi = \frac{E\pi^4 W^2}{a^2 b^2} \left(\cos^2 \frac{\pi x}{a} \cos^2 \frac{\pi y}{b} - \sin^2 \frac{\pi x}{a} \sin^2 \frac{\pi y}{b} \right)$$
$$= \frac{E\pi^4 W^2}{2a^2 b^2} \left(\cos \frac{2\pi x}{a} + \cos \frac{2\pi y}{b} \right)$$

and hence

$$\phi = \frac{p_1 y^2}{2} + \frac{p_2 x^2}{2} + \frac{EW^2}{32} \left(\frac{a^2}{b^2} \cos \frac{2\pi x}{a} + \frac{b^2}{a^2} \cos \frac{2\pi y}{b} \right) \tag{14.45}$$

where p_1 and p_2 are constants to be determined using the in-plane conditions [Eq. (14.43)].

Corresponding to the above ϕ, the in-plane force field is given by

$$N_x = p_1 h - \frac{Eh W^2 \pi^2}{8a^2} \cos \frac{2\pi y}{b}$$

$$N_y = p_2 h - \frac{Eh W^2 \pi^2}{8b^2} \cos \frac{2\pi x}{a}$$

$$N_{xy} = 0 \tag{14.46}$$

from which one can identify the physical significance of p_1 and p_2 as given by

$$\int_0^b N_x dy = p_1 bh = \text{Total normal force across any section } x = \text{constant}$$

$$\int_0^a N_y dx = p_2 bh = \text{Total normal force across any section } y = \text{constant} \tag{14.47}$$

Use of Eqs. (14.37) and (14.39) along with Eq. (14.46) yields $u_{o,x}$ and $v_{o,y}$ in terms of p_1, p_2 and W; integration of these with respect to x and y, respectively, and enforcement of zero u_o along $x = a/2$ and zero v_o along $y = b/2$ (i.e. assuming that the deformation of the simply supported plate is symmetric about its central lines) leads to

$$u_o = \left(\frac{p_1 - \mu p_2}{E} - \frac{W^2 \pi^2}{8a^2} \right) \left(x - \frac{a}{2} \right) - \frac{W^2 \pi a}{16} \sin \frac{2\pi x}{a} \left(\frac{1}{a^2} - \frac{\mu}{b^2} - \frac{1}{a^2} \cos \frac{2\pi y}{b} \right)$$

$$v_o = \left(\frac{p_2 - \mu p_1}{E} - \frac{W^2 \pi^2}{8b^2} \right) \left(y - \frac{b}{2} \right) - \frac{W^2 \pi b}{16} \sin \frac{2\pi y}{b} \left(\frac{1}{b^2} - \frac{\mu}{a^2} - \frac{1}{b^2} \cos \frac{2\pi x}{a} \right)$$

$$\tag{14.48}$$

Hence, the edge displacements are given by

$$u_o|_{x=a} = -u_o|_{x=0} = \left(\frac{p_1 - \mu p_2}{E} - \frac{W^2 \pi^2}{8a^2} \right) \frac{a}{2}$$

$$v_o|_{y=b} = -v_o|_{y=0} = \left(\frac{p_2 - \mu p_1}{E} - \frac{W^2 \pi^2}{8b^2} \right) \frac{b}{2} \tag{14.49}$$

which are constant along the corresponding edges.

If the edges are immovable as per Eq. (14.43a), then by equating the above displacements to zero, one gets

$$p_1 = \frac{E W^2 \pi^2}{8(1 - \mu^2)} \left(\frac{1}{a^2} + \frac{\mu}{b^2} \right), \quad p_2 = \frac{E W^2 \pi^2}{8(1 - \mu^2)} \left(\frac{1}{b^2} + \frac{\mu}{a^2} \right) \tag{14.50a}$$

If the edges are movable as per Eq. (14.43b) with the net normal force zero, then

$$p_1 = p_2 = 0 \tag{14.50b}$$

The edge conditions of zero in-plane shear are satisfied automatically by the chosen stress function; in fact N_{xy} is zero everywhere in the plate [see Eq. (14.46)].

With the values for p_1 and p_2 substituted in Eq. (14.45), the stress function ϕ is now expressed in terms of the unknown W alone. Use of this and the assumed sinusoidal variation for W as per Eq. (14.44) in the left-hand side of the transverse equilibrium equation (Eq. 14.41) provides a residual R which can be orthogonalized, as per Galerkin's method, as

$$\int\limits_0^a \int\limits_0^b R \sin \frac{\pi x}{a} \sin \frac{\pi y}{b} dy\, dx = 0$$

since the assumed function for w satisfies all the boundary conditions of Eq. (14.42).

This procedure yields the maximum deflection W in terms of the uniform load q_o as

$$q_o = \frac{\pi^6 DW}{16} \left(\frac{1}{a^2} + \frac{1}{b^2}\right)^2 \left[1 + k_{\mathrm{NL}} \left(\frac{W}{h}\right)^2\right] \qquad (14.51)$$

where k_{NL} is a parameter that quantifies the non-linearity and is given by

$$k_{\mathrm{NL}} = \frac{3(1 - \mu^2)}{4(a^2 + b^2)^2} \left[a^4 + 2\beta a^2 b^2 + (1 + 2\alpha)b^4\right] \qquad (14.52)$$

with

$$\left. \begin{aligned} \alpha &= \frac{1}{(1 - \mu^2)}\left(1 + \frac{\mu a^2}{b^2}\right) \\ \beta &= \frac{1}{(1 - \mu^2)}\left(\mu + \frac{a^2}{b^2}\right) \end{aligned} \right\} \text{ if the edges are immovable;}$$

$$\alpha = \beta = 0 \text{ if the edges are movable.}$$

Thus, the non-linearity is similar to that found in the case of cylindrical bending considered in Sect. 14.1.1, for which the non-linear parameter k_{NL} is 3. Here k_{NL} depends on a/b as well as μ, and of course, on whether the edges are immovable or movable; for $\mu = 0.316$ (the value for which exact results are available[5]), one gets $k_{\mathrm{NL}} = 1.325$ (immovable), 0.338 (movable) for a square plate, and $1.631, 0.459$, respectively, for $a/b = 2$.

The load–deflection plots for the square plate are shown in Fig. 14.6 along with

---

[5]S. Levy, Bending of rectangular plates with large deflections, NACA Report No. 737, 1942 (*available online for open access*).

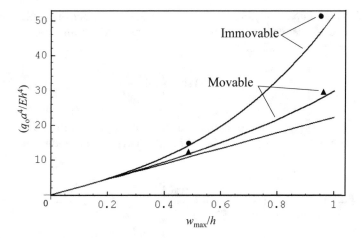

Fig. 14.6 Linear and non-linear load–deflection plots (● and ▲—Exact solution due to Levy)

the linear counterpart. To provide an idea of the accuracy of the above Galerkin solution, a few data points taken from Levy's report are also included in Fig. 14.6; they differ from the Galerkin solution by less than 5%.

From this figure, it is clear that departure from linearity occurs a little earlier for the immovable case as compared to the movable case as expected, and that the errors of the linear deflection results are unacceptably high when they become comparable to the thickness. As far as stresses are concerned, it can be shown that the linear theory over-estimates the maximum stress and hence leads to over-design, as was the case with the cylindrical bending problem discussed earlier.

We shall stop with this simple example, but point out that non-linear bending of plates of various shapes and boundary conditions has been studied exhaustively and carefully documented[6] in the form of graphs and tables for direct use by designers.

14.4   Non-linear Vibrations of a Rectangular Plate

Let us now consider free vibrations of the above simply supported rectangular plate with due account of the effect of mid-plane stretching; we shall do this by following the same steps as for the static problem, but now applied to the equation of motion as given by

$$DV^4 w = -\overline{\rho} w_{,tt} + h\phi_{,yy} w_{,xx} + h\phi_{,xx} w_{,yy} - 2h\phi_{,xy} w_{,xy} \qquad (14.53)$$

---

[6] See, for example:

C.Y. Chia, Non-linear Analysis of Plates, McGraw-Hill, 1980;

W.C. Young, R.G. Budynas, Roark's Formulas for Stress and Strain, McGraw-Hill, 2001.

along with the compatibility Eq. (14.40) which holds good when the in-plane inertias are neglected. As defined earlier in Chap. 6, $\bar{\rho}$ is the mass per unit area.

Seeking a one-term solution as

$$w = W(t) \sin \frac{\pi x}{a} \sin \frac{\pi y}{b} \tag{14.54}$$

the stress function turns out to be the same as in Eq. (14.45) with the values of p_1 and p_2 as in Eq. (14.50). Applying Galerkin's method to the equation of motion now yields

$$\left(\frac{W}{h}\right)_{,tt} + \frac{\pi^4 D}{\bar{\rho}} \left(\frac{1}{a^2} + \frac{1}{b^2}\right)^2 \left(\frac{W}{h}\right) \left[1 + k_{NL} \left(\frac{W}{h}\right)^2\right] = 0 \tag{14.55}$$

with k_{NL} as in Eq. (14.52). This equation with a cubic restoring force term is often referred to as a Duffing-type equation.

This equation has an exact solution; if the initial conditions correspond to a non-zero displacement $W(0) = W_o$ and zero velocity, it is given by

$$\frac{W}{h} = \frac{W_o}{h} \mathrm{cn}(\omega t, \bar{k}) \tag{14.56}$$

where $\omega = \omega_{\mathrm{lin}} \sqrt{1 + k_{NL}(W_o/h)^2}$
and $\bar{k}^2 = \frac{k_{NL}(W_o/h)^2}{2[1+k_{NL}(W_o/h)^2]}$; $\omega_{\mathrm{lin}}^2 = \frac{\pi^4 D}{\bar{\rho}}\left(\frac{1}{a^2} + \frac{1}{b^2}\right)^2$.

The function $\mathrm{cn}(\omega t, \bar{k})$ is known as Jacobi's elliptic function which reduces to the simple trigonometric cosine function when its second argument—the so-called modulus $\bar{k}$—becomes zero, i.e. when the deflection W_o is very small compared to h; this is the familiar harmonic vibration corresponding to small amplitudes occurring at the linear frequency ω_{lin}. When the modulus is not small, the response corresponding to the elliptic function is a periodic function of amplitude W_o/h, similar to the simple cosine function but not exactly the same; the period of this function is given by

$$T_{NL} = \frac{4K(\bar{k})}{\omega} \tag{14.57}$$

where $K(\bar{k})$ is the complete elliptic integral of the first kind. This period is shorter than the linear period T_{lin} given by $(2\pi/\omega_{\mathrm{lin}})$.

To get a feel for this non-linearity, let us consider a square aluminium plate ($E = 70$ GPa, $\mu = 0.3$) of 300 mm side and 2 mm thickness. Let the simply supported edges be of the immovable type for which k_{NL} can be calculated from Eq. (14.52) as 1.316. The linear frequency turns out to be 688.75 rad/s, i.e. $T_{\mathrm{lin}} = 0.0091$ s. Using *Mathematica*, the non-linear period T_{NL} can be calculated using Eq. (14.57)

Fig. 14.7 Non-linear versus linear response

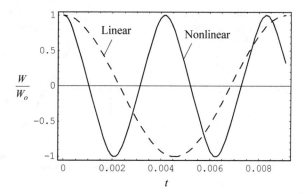

Fig. 14.8 Non-linear period versus amplitude for a square plate

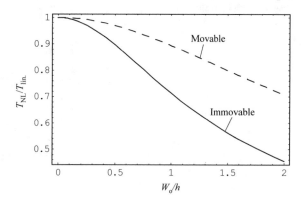

for any W_o/h; for $W_o/h = 0.5, 1.0$ and 2.0, T_{NL} turns out to be 0.0082, 0.0065 and 0.0042 s. Thus, due to the hardening non-linearity, the restoring action speeds up the oscillations. This is clear from the response plots shown in Fig. 14.7 for $W_o/h = 2.0$.

The ratio of the periods is plotted against the amplitude-to-thickness ratio in Fig. 14.8 for both immovable and movable edge conditions. Further, to differentiate the elliptic function from the simple cosine function, they are plotted together in Fig. 14.9 taking the period to be the same.

Results similar to the above are available for different boundary conditions and different plate aspect ratios, obtained using a variety of methods.[7] Apart from the dependence of the free vibration frequencies on the amplitude and the fact that the free vibration response is no more a harmonic function, the behaviour of a non-linear dynamic system differs from that of a linear system in several ways. A discussion of such non-linear phenomena is appropriate at this stage, but this cannot be done at great length here. We shall devote just a few paragraphs to kindle the reader's

---

[7]C.Y. Chia, Non-linear Analysis of Plates, McGraw-Hill, 1980.

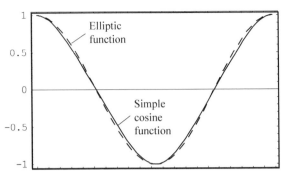

Fig. 14.9 Simple cosine function versus elliptic function of the same period

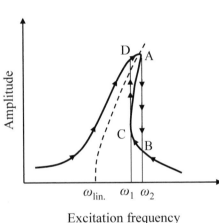

Fig. 14.10 Steady-state response of a hardening system

interest regarding this exciting subject, which has been dealt with more elaborately in several books.[8]

Consider the forced vibration of a single degree-of-freedom hardening system due to a harmonic exciting force. When the exciting frequency is close to the linear natural frequency leading to an increase of amplitudes due to resonance, the natural frequency of the system changes (actually increases), thus limiting the amplitude to a finite value. If the exciting frequency is increased a little making it closer to the new natural frequency, the amplitude increases further due to resonance but settles down once again to a finite value. However, such an increase of the steady-state amplitude with increase of the exciting frequency does not occur indefinitely but only till a particular exciting frequency. If one solves the governing equation for such steady-state or stationary solutions, then the resulting frequency response curve, for a system with some damping, is as shown in Fig. 14.10; the free vibration amplitude versus frequency variation is shown as a dotted line in this figure and is usually referred to as the backbone curve. The most important feature that has to be noted here, in contrast with the behaviour of a linear system, is that there may be multiple stationary

---

[8]See, for example: J.J. Thomsen, Vibrations and Stability, Springer, 2003.

solutions for the same excitation frequency—this occurs in the range (ω_1, ω_2) where there are three possible values of the amplitude. Of these three, only one is observed in an experiment, and this depends on how the experiment is carried out, as explained below.

If one increases the exciting frequency from a small value, then the stationary amplitude increases continuously till it reaches the maximum (point A) at ω_2 as shown by the arrows in Fig. 14.10. On the other hand, if the frequency sweep is carried out by decreasing from a large exciting frequency, the amplitude variation is along BC.

At C, if the exciting frequency is further lowered to a little below ω_1, the stationary amplitude undergoes a jump to D, this change occurring during some transient oscillations till the system achieves a steady state once again. Similarly, there is a jump from A to B in the increasing frequency sweep. The intermediate values of amplitudes, along AC, do not occur in practice.

Besides the primary resonance described above, the non-linear system may exhibit large amplitude resonant oscillations at a frequency close to $\omega_{lin.}$ when the exciting frequency is a fraction or a multiple of $\omega_{lin.}$. For the Duffing-type oscillator, such severe oscillations would occur when the exciting frequency is $\omega_{lin.}/3$ (called *superharmonic resonance*) and also when it is $3\omega_{lin.}$ (called *sub-harmonic resonance*).

For multi-degree-of-freedom or continuous systems, the forced vibration response becomes much more complicated when there is more than one principal mode with the same linear natural frequency or when there is a simple linear algebraic relationship between three or more natural frequencies (referred to as the cases of *internal resonance*). A discussion of such problems, involving plates, can be found in the literature.[9]

As mentioned earlier, the discussion of non-linear vibrations here is quite brief; for a more complete understanding of non-linear behaviour of isotropic and orthotropic plates as well as several methods employed for non-linear analysis, one should consult appropriate review articles.[10]

Summary Geometrically non-linear behaviour of plates has been briefly discussed here with reference to problems of static flexure and vibrations. The important effects of non-linearity have been highlighted, and several references have been cited for further study.

---

[9]See, for example:

G. Anlas, O. Elbeyli, Nonlinear vibrations of a simply supported rectangular metallic plate subjected to transverse harmonic excitation in the presence of a one-to-one internal resonance, Nonlinear Dynamics, 30, 2002, 1-28;

S. Sridhar, D.T. Mook, A.H. Nayfeh, Non-linear resonances in the forced responses of plates, Part 1: Symmetric responses of circular plates, Jl. of Sound and Vibration, 41, 1975, 359–373.

[10]See, for example: M. Sathyamoorthy, Nonlinear vibrations of plates: An update of recent research developments, Applied Mechanics Reviews, 49(10), Part 2, 1996, S55–S62.

Chapter 15
Post-buckling Behaviour

Unlike columns, rectangular plates with supported sides can carry significant loads beyond the initial buckling load, and this reserve strength often needs to be utilized in order to arrive at an efficient low-weight design. The behaviour after the onset of buckling cannot be captured using linear analysis as carried out earlier in Chap. 7, but requires the use of the moderately large deformation theory put forth in the last chapter. Such post-buckling analysis is the topic of this chapter; for better understanding, the simpler case of a column is discussed first.

15.1 Post-buckling of a Column

Let us first consider the familiar problem of a simply supported slender perfect column of length L (Fig. 15.1), noting that the conclusions are also applicable to cylindrical bending of plate strips with unsupported sides.

Euler's analysis of this problem is based on the linear expression $w_{,xx}$ for the curvature and yields the lowest critical load as $\frac{\pi^2 EI}{L^2}$, which may be defined as the load at which an adjacent bent equilibrium configuration is possible besides the undisturbed straight configuration. However, this analysis does not throw any light on the actual transverse deflections corresponding to this Euler load and on whether higher loads are possible at all.

To answer such questions, one has to use the exact non-linear curvature expression [see Eqs. (14.15) and (14.16)]; such a large deformation analysis, referred to as the *elastica problem*, yields a specific deformed position corresponding to each value

© The Author(s) 2021

K. Bhaskar and T. K. Varadan, *Plates*,

https://doi.org/10.1007/978-3-030-69424-1_15

Fig. 15.1 Elastica problem

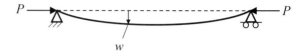

Fig. 15.2 Post-buckling
deflections of a column

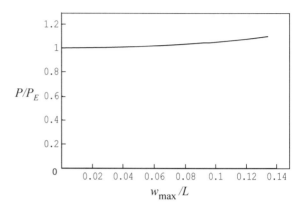

of the compressive load P. The final solution from such an analysis is in terms of
elliptic integrals[1] and is not presented here.

A simplified expression for the central deflection, sufficient for our discussion
and reasonably accurate for not-very-large deflections, is given by

$$w_{max} = \frac{2L}{\pi} \sqrt{\frac{P_E}{P} \left(\sqrt{\frac{P}{P_E}} - 1 \right)} \qquad (15.1)$$

where P_E is the Euler load. This equation yields an imaginary value for the deflection
when $P < P_E$, which simply means that there is no bent equilibrium position for sub-
critical loads. As P is increased beyond P_E, specific values for w_{max} are obtained
as, for example, $0.014L$, $0.045L$, $0.063L$ and $0.13L$, respectively, for $P/P_E = 1.001$,
1.01, 1.02 and 1.1. Thus, excessive bending deflections are associated with loads
slightly in excess of the Euler load; this is shown graphically in Fig. 15.2.

Considering a column of circular cross section of diameter d and length
$L = 30d$, one can now compare the maximum stresses at different values of P. At
$P = P_E$, the maximum stress, uniform over the entire cross section of the as yet unbent
column, is $\frac{4P_E}{\pi d^2}$. At $P = 1.001 P_E$, the maximum stress, calculated as the sum of such a
uniform stress $\frac{4P}{\pi d^2}$ and the bending stress at the extreme fibre $\frac{(Pw_{max})(d/2)}{(\pi d^4/64)}$, is obtained
as $\frac{17.46 P_E}{\pi d^2}$. Similar calculations yield $\frac{47.67 P_E}{\pi d^2}$ at $P = 1.01 P_E$. Thus, the maximum
stress increases rapidly as P is increased beyond the Euler load and yielding of the
column would begin at a load very close to P_E.

---

[1] See: S.P. Timoshenko, J.M. Gere, Theory of Elastic Stability, McGraw-Hill, 1963.

Hence, for all practical purposes, the post-buckling reserve strength of a column should be considered negligibly small and P_E itself should be taken as the ultimate load.

15.2   Post-buckling of a Rectangular Plate

For this discussion, let us consider a specific problem amenable to a fairly simple solution—the case of a rectangular plate $(0 \leq x \leq a, 0 \leq y \leq b)$ of thickness h with bending-resistant, straight, movable simply supported edges (defined as in Sect. 14.3), and with the in-plane tangential displacements at the edges freely permitted and the corresponding shear forces zero. The plate is subjected to uniaxial compression in the x-direction by loads applied on the edges $x = 0, a$; the magnitude of these loads can be expressed in terms of the average pressure p_x as $p_x hb$. The problem is to seek bent equilibrium configurations of this compressed plate; for this purpose, we shall assume that the deflections are not too large compared to the thickness of the plate so that the moderately large deformation theory presented in the earlier chapter can be used.

Thus, we seek a solution of the following equations:

$$\nabla^4 \varphi = Eh(w_{,xy}^2 - w_{,xx}w_{,yy}) \tag{15.2a}$$

$$D\nabla^4 w = h\varphi_{,yy}w_{,xx} + h\varphi_{,xx}w_{,yy} - 2h\varphi_{,xy}w_{,xy} \tag{15.2b}$$

$$w = w_{,xx} = 0 \ \text{ at } x = 0, a$$
$$w = w_{,yy} = 0 \ \text{ at } y = 0, b \tag{15.3a}$$

$$u = \text{constant along } y, \text{ with } \int_{y=0}^{b} N_x dy = -p_x hb, \quad N_{xy} = 0 \text{ at } x = 0, a$$

$$v = \text{constant along } x, \text{ with } \int_{x=0}^{a} N_y dx = N_{xy} = 0 \text{ at } y = 0, b \tag{15.3b}$$

We shall employ the method originally presented by Marguerre[2] starting with an approximation of the buckled shape $w(x, y)$, followed by the determination of the corresponding Airy stress function using Eq. (15.2a), and finally an approximate solution of Eq. (15.2b). In fact, the static flexure problem of Sect. 14.3 was solved

---

[2]K. Marguerre, The apparent width of the plate in compression, NASA TM No. 833, July 1937 (*available online for open access*).

by the same approach, and hence a close correspondence with that section can be seen here.

Assuming that the buckled shape can be adequately captured using a one-term approximation satisfying Eq. (15.3a) as

$$w = W \sin \frac{\pi x}{a} \sin \frac{\pi y}{b} \tag{15.4}$$

the stress function is obtained as

$$\varphi = \frac{p_1 y^2}{2} + \frac{p_2 x^2}{2} + \frac{EW^2}{32} \left(\frac{a^2}{b^2} \cos \frac{2\pi x}{a} + \frac{b^2}{a^2} \cos \frac{2\pi y}{b} \right) \tag{15.5}$$

Use of Eq. (15.3b) yields the constants p_1 and p_2 as

$$p_1 = -p_x, \quad p_2 = 0 \tag{15.6}$$

Equation (15.2b) is now satisfied in a weighted-residual sense using Galerkin's method since the assumed function for w satisfies all the boundary conditions. This yields

$$p_x = \frac{\pi^2 D (a^2 + b^2)^2}{ha^2 b^4} + \frac{\pi^2 EW^2(a^4 + b^4)}{16 a^2 b^4} \tag{15.7}$$

At this stage, let us specialize for a square plate by putting $a = b$, noting that this is not too restrictive since a rectangular plate buckles into nearly square panels separated by nodal lines and hence its post-buckling behaviour is qualitatively the same as that of a square plate.

For the square plate, we have

$$p_x = \frac{4\pi^2 D}{hb^2} + \frac{\pi^2 EW^2}{8b^2} \tag{15.8}$$

where the first term can be identified as the critical pressure ($|N_{xcr}|/h$) obtained using linear stability analysis (Chap. 7). Denoting this by p_{cr}, one then gets

$$W \propto \sqrt{p_x - p_{cr}} \tag{15.9}$$

which clearly shows that no buckling is possible for sub-critical pressures.

Using the definition of D, the actual expression for the maximum deflection W can be obtained as

$$W = h \sqrt{\frac{8}{3(1 - \mu^2)} \left(\frac{p_x}{p_{cr}} - 1 \right)} \tag{15.10}$$

Fig. 15.3 Post-buckling deflections of a plate

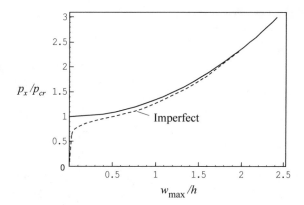

which is an important equation that should be contrasted with its counterpart for the column (Eq. (15.1)).

The difference between the column and the plate with supported sides is that while the deflections for the former grow rapidly to values of the order of L as the critical load is exceeded, they grow quite slowly for the latter and remain comparable to the thickness h even for fairly large post-critical loads. For example, for a material with $\mu = 0.3$, Eq. (15.10) yields $W = 0.54\,h$, $1.21\,h$, $1.71\,h$ and $2.42\,h$, respectively, for $p_x/p_{cr} = 1.1$, 1.5, 2 and 3. The load-transverse deflection curve is as shown in Fig. 15.3, where one should note that the rate of increase of the deflections decreases as the load p_x is increased beyond p_{cr}, or in other words, the plate becomes stiffer.

To explain why the plate is capable of carrying a significantly larger load beyond the critical value, one has to look at the membrane stress distribution. Prior to buckling, the applied load is resisted uniformly by the plate with the internal stress field

$$\sigma_x = -p_x, \quad \sigma_y = \tau_{xy} = 0 \tag{15.11}$$

as shown in Fig. 15.4.

Fig. 15.4 Stress field for sub-critical loads

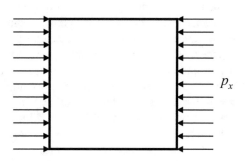

Fig. 15.5 Membrane
stresses after buckling

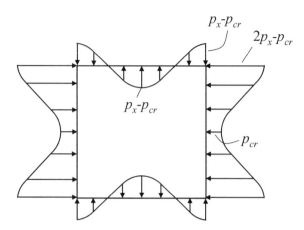

Beyond $p_x = p_{cr}$, corresponding to the stress function in Eq. (15.5), we have

$$\sigma_x = p_1 - \frac{\pi^2 W^2 E}{8b^2} \cos \frac{2\pi y}{b} = -\left(p_x + (p_x - p_{cr}) \cos \frac{2\pi y}{b}\right)$$

$$\sigma_y = -\frac{\pi^2 W^2 E}{8b^2} \cos \frac{2\pi x}{a} = -\left((p_x - p_{cr}) \cos \frac{2\pi x}{a}\right)$$

$$\tau_{xy} = 0 \tag{15.12}$$

Considering σ_x first, one can see that it is compressive for all y with a larger value at the supported sides ($y = 0, b$), this value increasing with p_x as $(2p_x - p_{cr})$; at the centre $y = b/2$, it is always p_{cr} irrespective of the actual load p_x.

The stress σ_y should have an average value of zero corresponding to the movable sides $y = 0, b$. However, the actual distribution is in the form of a cosine function with compressive values at the ends $x = 0, a$ and a maximum tensile value at the centre $x = a/2$, both increasing with p_x and equal to $(p_x - p_{cr})$. The variations are as shown in Fig. 15.5.

From the above, one can summarize that the post-buckling reserve strength of the plate with supported sides is due to the following reasons. While the applied load is taken up by the entire width of the plate prior to buckling, any additional load after buckling is taken up by the portions of the plate close to the supported sides, while the central portion which has bulged out remains stressed only at p_{cr}. Further, transverse stresses σ_y arise with tensile values at the centre of the plate in between the loaded ends. Thus, while the central portion bulges out as the plate buckles at p_{cr}, the bulge increases only gradually with increasing additional load because of the redistribution of σ_x and the stabilizing influence of the tensile σ_y. The final failure of the plate occurs when the maximum compressive membrane stress σ_x at the sides $y = 0, b$ exceeds the allowable value, which may be the yield strength of the plate material or one corresponding to buckling failure of the members supporting the sides; this final value of the total load (hbp_x) is the ultimate load of the plate. At this load, the central bulged portion is often quite under-stressed, well within the elastic limit, even if one considers the bending stresses besides the membrane stresses.

Fig. 15.6 Post-buckling
behaviour under shear

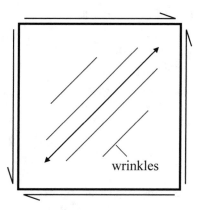

The above conclusions are based on a simple one-term approximation for the buckled shape which is reasonably accurate for plates with the ultimate load less than about 3 times the critical load; for steel plates, this amounts to the restriction that the plate thickness ratio b/h be less than about 100. If the plate is thinner such that the ratio of the ultimate load to the critical load is much larger, the above described post-buckling behaviour is valid soon after p_{cr} is exceeded, but not for higher loads because the portions of the plate close to the sides tend to buckle as independent long strips into nearly square panels. Thus, the buckled shape of the whole plate is more complicated than can be described by just one term as in Eq. (15.4); to overcome this deficiency, a three-term solution was also presented by Marguerre.[3] Such a more complicated solution shows that the one-term solution is non-conservative for very thin plates because it over-predicts the ultimate strength, and hence a multi-term approach is essential.

In the above analysis, it has been assumed that the plate is initially perfectly flat; in practice, because of imperfections, the plate starts bending as soon as the compressive load is applied and the load–deflection behaviour is as indicated by the dotted line in Fig. 15.3.[4] It is also appropriate to point out that the above conclusions, deduced with reference to a particular simply supported plate, are qualitatively applicable for plates with other boundary conditions such as clamped edges.

In the case of pure shear loading, the post-buckling strength of a square plate arises in a slightly different manner; here as the plate is compressed in the direction of one diagonal and elongated in the direction of the other, it buckles and becomes ineffective in the former direction, but continues to carry higher stresses in the latter (Fig. 15.6). This is the basis for the tension field theory[5] applicable to thin-walled

---

[3] K. Marguerre, The apparent width of the plate in compression, NACA TM No. 833, 1937 (*available online for open access*).

[4] P.C. Hu, E.E. Lundquist, S.B. Batdorf, Effect of small deviations from flatness on effective width and buckling of plates in compression, NACA TN 1124, 1946 (*available online for open access*).

[5] T.H.G. Megson, Aircraft Structures for Engineering Students, 2007, Elsevier.

Fig. 15.7 Effective width

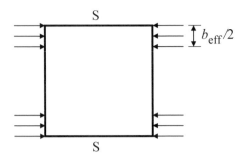

beams involving a thin stiffened web between top and bottom flanges, typically used in the construction of airplane wings.

It is necessary to note that the literature on post-buckling of plates is quite exhaustive and covers the use of several approximate analytical and numerical approaches such as the finite element method which have been recognized as powerful tools for non-linear analysis. However, the simple analytical approach presented above not only serves to highlight the salient features, but also remains a viable approach even for more complicated problems as exemplified by its application to imperfect composite plate problems recently and demonstration of its validity by comparison with some costly computational approaches.[6]

15.3 Effective Width

The discussion of post-buckling behaviour of a plate would be incomplete without the mention of the concept of effective width. While non-linear analysis using the moderately large deformation theory remains a rigorous approach, it is not always convenient for predicting the ultimate load capacity of a plate. Simpler alternatives, based on intuitive assumptions, are necessary, and one such approach, originally put forth by von Karman in 1932, is to assume that the central portion of the uniaxially compressed plate is totally ineffective in carrying stress, while the two strips close to the supported sides resist the applied load uniformly (Fig. 15.7). The total width of the two strips is defined as the effective width b_{eff}.

If σ_{all} is the maximum allowable compressive stress, then the ultimate load of the plate is given by

$$P_u = b_{eff} h \sigma_{all} \tag{15.13}$$

[6]C. Mittelstedt, K. Schroder, Post-buckling of compressively loaded imperfect composite plates: closed-form approximate solutions, International Jl. of Structural Stability and Dynamics, 10, 2010, 761–778.

The effective width, according to von Karman, can be calculated by assuming that P_u corresponds to the critical load of a long simply supported plate of width b_{eff}, i.e.

$$P_u = b_{eff} \left(\frac{4\pi^2 D}{b_{eff}^2} \right) \tag{15.14}$$

Comparing the above expressions, one gets

$$b_{eff} = 2\pi \sqrt{\frac{D}{h\sigma_{all}}}$$

$$= 2\pi h \sqrt{\frac{E}{12(1 - \mu^2)\sigma_{all}}} = 1.9h \sqrt{\frac{E}{\sigma_{all}}} \quad \text{for } \mu = 0.3 \tag{15.15}$$

This formula is referred to as von Karman's effective width formula. For better correlation with experimental data and to arrive at a conservative estimate of the ultimate load, the coefficient is often taken as 1.7 instead of 1.9.

The above method can be repeated when one side is free and the other supported by noting that in this case only the strip close to the supported side carries the load; the final result is a similar equation with 0.62 in place of 1.9. Again, by comparison with experiments, a value of 0.60 is commonly used.

It should be noted that several other formulae have been suggested for the effective width and have been documented and compared in the literature,[7] though von Karman's equations still remain popular primarily because of their simplicity.

Summary

A very brief introduction to the post-buckling behaviour of a plate has been presented here, with just enough details to highlight the importance of such a study. No complicating factors such as material inhomogeneity or orthotropy or inelastic behaviour have been considered. For such topics, the interested reader is encouraged to refer to specialized monographs as well as the exhaustive open literature on post-buckling.

---

[7] see, for example: T.V. Galambos, Guide to Stability Design Criteria for Metal Structures, Wiley, 1998.

Appendix
Solutions for Problems–Chapter 3

Problem 1
The CPT solution is obtained by successive integration of Eq. (3.1) and the use of the clamped edge conditions, as.

$$Dw = q_o \left(\frac{a}{\pi}\right)^4 \sin \frac{\pi x}{a} + q_o \left(\frac{a}{\pi}\right)^3 (x^2 - x)$$

The lines of contraflexure, corresponding to $w_{,xx} = 0$, are at $x = 0.2197a$ and $0.7803a$, i.e. $0.5606a$ apart. Hence, this portion has a span-to-thickness ratio equal to about 56% of the overall value. This result does depend on the load distribution—for instance, one can verify that the lines of contraflexure are $0.5774a$ apart for the case of a uniformly distributed load.

Problem 2
The maximum bending stress σ_x turns out to be equal for both the theories—this is because in each case it is given by $(6M_x/h^2)$ and M_x simply depends on the load distribution.

The strain energies by the two theories are related as

$$\frac{U_{e\,NEW}}{U_{e\,CPT}} = \frac{(1 - 2\mu)}{(1 - \mu)^2}$$

This result can be obtained by direct calculation of the strain energy, or by using the fact that while $\sigma_x(x, z)$ is the same in both the theories, $\varepsilon_x(x, z)$ by the new theory is under-predicted by a factor of $(1-\mu)^2/(1 - 2\mu)$, i.e. 1.225 for $\mu = 0.3$.

© The Author(s) 2021
K. Bhaskar and T. K. Varadan, *Plates*,
https://doi.org/10.1007/978-3-030-69424-1

Appendix
Solutions for Problems–Chapter 4

Problem 1
The non-zero Fourier coefficients are given by

(a) $q_{mn} = \frac{32q_o(-1)^{\frac{m-1}{2}}}{m^2 n\pi^3}$ for $m, n = 1, 3, 5, \ldots$

(b) $q_{mn} = \frac{4q_o(-1)^{\frac{m-1}{2}}}{m\pi}$ for $m = n = 1, 3, 5, \ldots$

Problem 2
The applied load is equivalent to a uniform load q_o over the square patch of side $3a/4$ minus the same load q_o over the square patch of side $a/4$. The Fourier coefficients for a central patch load of intensity q_o (on a square of side u) are as given in Example 4.2 with (P/u^2) replaced by q_o. These can be used straightaway here—proceeding thus, one obtains the coefficients of the series for the deflection as

$$W_{mn} = \frac{16q_o a^4 (-1)^{\frac{m+n}{2}-1}}{\pi^6 Dmn(m^2+n^2)^2}\left(\sin\frac{3m\pi}{8}\sin\frac{3n\pi}{8} - \sin\frac{m\pi}{8}\sin\frac{n\pi}{8}\right)$$

for odd m and n.

Problem 3
The upward shear force along either $x = 0$ or $x = a$ is given by

$$|Q_x| = \frac{q_o}{a\pi\left(\frac{1}{a^2}+\frac{1}{b^2}\right)}\sin\frac{\pi y}{b}$$

while that along either $y = 0$ or $y = b$ is

$$|Q_y| = \frac{q_o}{b\pi\left(\frac{1}{a^2}+\frac{1}{b^2}\right)}\sin\frac{\pi x}{a}$$

© The Author(s) 2021
K. Bhaskar and T. K. Varadan, *Plates*,
https://doi.org/10.1007/978-3-030-69424-1

Integrating these forces along the corresponding edges, the total vertical reaction turns out to be $(4q_oab/\pi^2)$ which is equal to the integral of the load q over the area of the plate.

Problem 6
The non-zero Fourier coefficients of the given load are given by

$$q_{m1} = \frac{-8q_o}{\pi m(m^2 - 4)}\text{for odd } m.$$

These can be used to find out the deflection.

Problem 7
For a plate occupying the region $0 \le x \le a$, $0 \le y \le b$, the load can be taken as $q(x, y) = q_o x/a$ with the following non-zero Fourier coefficients.

$$q_{mn} = \frac{8q_o(-1)^{m+1}}{mn\pi^2}\text{for } m = 1, 2, \ldots \text{ and } n = 1, 3, 5, \ldots$$

The deflections, moments and shear forces can then be obtained as in Example 4.1.

Problem 8
(a) The solution is straightforward; the Fourier series for the patch load is as in Example 4.2 and that for the uniform load due to self-weight is as in Example 4.1. The corresponding one-term solutions for the maximum deflection are

$W_{11} = 0.091$ mm due to the patch load alone
and 0.695 mm due to self-weight alone.

(b) Noting that D is proportional to h^3 and the intensity of the self-weight to h, one can see that w_{max} due to self-weight alone is proportional to $1/h^2$. Thus, a change of thickness from 5 to 4 mm changes w_{max} to 1.09 mm from the earlier value of 0.695 mm.

Problem 10

(a) The line load is first converted into a uniform pressure of intensity P/u acting on a strip of width u. The deflection corresponding to this distributed load is found out by Navier's approach, and then the solution for the line load is obtained for the limiting case of $u \to 0$. The corresponding non-zero Fourier coefficients turn out to be

$$W_{mn} = \frac{8P \sin \frac{m\pi}{2} \sin \frac{n\pi}{2} \sin \frac{n\pi c}{2b}}{a\pi^5 Dn\left(\frac{m^2}{a^2} + \frac{n^2}{b^2}\right)^2}$$

for odd m and n.

(b)   Taking the first four non-zero terms, i.e. m and n up to 3 each, one obtains the
required thickness to be 2.59 mm for a maximum deflection of 1 mm.

Problem 11

The central deflection of a square plate ($0 \leq x, y \leq a$) due to a load P, acting on a
square patch of side u centred at an arbitrary point (ξ, η), is first found out. Taking
limits as $u \to 0$, one gets the central deflection due to a concentrated load P at (ξ, η)
as

$$w_c = \frac{Pa^2}{\pi^4 D} \sin \frac{\pi \xi}{a} \sin \frac{\pi \eta}{a}.$$

For the given problem, due to symmetry of the applied loads, one gets

$$w_c = 4[w_c \text{ due to } P \text{ at}(0.25, 0.25) + w_c \text{ due to } P \text{ at}(0.5, 0.25)].$$

Proceeding thus, one gets $P = 9.98$ N.

Problem 12

Adopting the procedure employed in Problem 10 above, one gets the non-zero Fourier
coefficients for the deflection to be:

For a central horizontal line load (P_1 per metre):

$$W_{mn} = \frac{8P_1 \sin \frac{n\pi}{2}}{\pi^5 Dbm \left(\frac{m^2}{a^2} + \frac{n^2}{b^2}\right)^2} \text{ for odd } m.$$

For a central vertical line load (P_2 per metre):

$$W_{mn} = \frac{8P_2 \sin \frac{m\pi}{2}}{\pi^5 Dan \left(\frac{m^2}{a^2} + \frac{n^2}{b^2}\right)^2} \text{ for odd } n.$$

Using these and the one-term approach, one gets the required answer to be
$D = 184.04$ Nm.

For the same total load if distributed uniformly over the plate, one can use the
result of Example 4.1 to obtain $D = 117.16$ Nm.

Problem 13

This is solved using Eq. (4.12) to yield

$$k_F = \frac{4D\pi^4}{a^4}.$$

Problem 15

(a) Following Levy's methodology, one gets

$$w_1 = \frac{q_o a^4}{m^4 \pi^4 D} \sin \frac{m\pi x}{a}$$

$$w_2 = \left(A_m \cosh \frac{m\pi y}{a} + B_m \frac{m\pi y}{a} \sinh \frac{m\pi y}{a} \right) \sin \frac{m\pi x}{a}$$

where

$$A_m = -\frac{q_o a^4}{m^4 \pi^4 D} \left(\frac{\sinh p_m + p_m \cosh p_m}{p_m + \sinh p_m \cosh p_m} \right)$$

$$B_m = \frac{q_o a^4}{m^4 \pi^4 D} \left(\frac{\sinh p_m}{p_m + \sinh p_m \cosh p_m} \right)$$

(b) The points at which the deflection or the normal stresses reach their maximum values are easily identified. However, that for the maximum shear stress has to be found by locating the point of maximum M_{xy} using $M_{xy,y} = 0$.

The final answers are:

$$|w(a/8, 0)| = 3.901 \times 10^{-5} q_o a^4 / D$$
$$|\sigma_x(a/8, 0, h/2)| = 0.03719 q_o a^2 / h^2$$
$$|\sigma_y(a/8, a/2, h/2)| = 0.03799 q_o a^2 / h^2$$
$$|\tau_{xy}(0, 0.42a, h/2)| = 0.00978 q_o a^2 / h^2$$

Problem 16

(a) The solution is the same as in Problem 15(a) except that

$$A_m = -\frac{q_o a^4}{2 m^4 \pi^4 D} \left(\frac{2 + p_m \tanh p_m}{\cosh p_m} \right) \qquad B_m = \frac{q_o a^4}{2 m^4 \pi^4 D \cosh p_m}$$

(b) Again, the solution is the same as in Problem 15(a) except that

$$A_m = \frac{q_o a^4}{m^4 \pi^4 D} \left(\frac{\mu(1+\mu)\sinh p_m - \mu(1-\mu)p_m \cosh p_m}{(3+\mu)(1-\mu)\sinh p_m \cosh p_m - (1-\mu)^2 p_m} \right)$$

$$B_m = \frac{q_o a^4}{m^4 \pi^4 D} \left(\frac{\mu(1-\mu)\sinh p_m}{(3+\mu)(1-\mu)\sinh p_m \cosh p_m - (1-\mu)^2 p_m} \right)$$

As b/a increases, both A_m and B_m decrease rapidly for all the three cases; thus, for very high b/a (>10), w_2 becomes negligibly small compared to w_1. This indicates that as the plate becomes longer and longer in the y-direction, the boundary conditions at $y = -\pm b/2$ do not matter since the problem reduces to one of cylindrical bending.

Problem 17

The solution for w_1 and w_2 is the same as in Example 4.3 except that

$$A_m = \frac{4q_o a^4}{m^5 \pi^5 D} \left(\frac{\mu(1+\mu)\sinh p_m - \mu(1-\mu)p_m \cosh p_m}{(3+\mu)(1-\mu)\sinh p_m \cosh p_m - (1-\mu)^2 p_m} \right)$$

$$B_m = \frac{4q_o a^4}{m^5 \pi^5 D} \left(\frac{\mu(1-\mu)\sinh p_m}{(3+\mu)(1-\mu)\sinh p_m \cosh p_m - (1-\mu)^2 p_m} \right)$$

Problem 18

The solution for w_1 is the same as in Example 4.3. For w_2, only odd-m terms will be present but with all the four constants A_m to D_m (see Eqs. (4.20) and (4.23)). These constants are obtained from the following equations.

$$A_m \cosh p_m + B_m p_m \sinh p_m + C_m \sinh p_m$$
$$+ D_m p_m \cosh p_m + \frac{4q_o a^4}{\pi^5 m^5 D} = 0$$
$$A_m \sinh p_m + B_m(\sinh p_m + p_m \cosh p_m)$$
$$+ C_m \cosh p_m + D_m(\cosh p_m + p_m \sinh p_m) = 0$$
$$A_m(1-\mu)\cosh p_m + B_m[2\cosh p_m + (1-\mu)p_m \sinh p_m]$$
$$- C_m(1-\mu)\sinh p_m - D_m[2\sinh p_m$$
$$+ (1-\mu)p_m \cosh p_m] - \frac{4\mu q_o a^4}{\pi^5 m^5 D} = 0$$
$$A_m(1-\mu)\sinh p_m - B_m[(1+\mu)\sinh p_m$$
$$- (1-\mu)p_m \cosh p_m] - C_m(1-\mu)\cosh p_m$$
$$+ D_m[(1+\mu)\cosh p_m - (1-\mu)p_m \sinh p_m] = 0$$

For any specific data for the geometric and material parameters and for each value of m, the above equations can be solved easily using Cramer's rule.

Problem 19

The load can be expressed in a Fourier sine series as

$$q = \sum_m q_m \sin \frac{m\pi x}{a} \quad \text{where } q_m = \frac{2q_o}{m\pi}(1 - \cos \frac{m\pi}{2})$$

The corresponding solution for w_1 is given by

$$w_1 = \sum_m \frac{q_m}{D} \left(\frac{a}{m\pi} \right)^4 \sin \frac{m\pi x}{a}$$

Only A_m and B_m appear in w_2, and they are given by

$$A_m = -\frac{q_m a^4}{2m^4\pi^4 D}\left(\frac{p_m \tanh p_m + 2}{\cosh p_m}\right); \quad B_m = \frac{q_m a^4}{2m^4\pi^4 D \cosh p_m}$$

The maximum value of τ_{xz} occurs at $(0, 0, 0)$ and is obtained as

$$\sum_m \frac{3q_m a}{2hm\pi}(1 - \operatorname{sech} p_m)$$

Problem 20

The solution for w_1 corresponding to the given load ($q = q_o x/a$, with the origin at the top-left corner of the plate) is

$$w_1 = \frac{q_o x}{360 a D}(3x^4 - 10a^2 x^2 + 7a^4)$$

$$= \sum_m \frac{2q_o a^4(-1)^{m+1}}{m^5\pi^5 D}\sin\frac{m\pi x}{a}$$

Due to lack of symmetry of the boundary conditions about $x = 0$, all the constants A_m to D_m are present in w_2. The corresponding algebraic equations are best solved numerically for each value of m, yielding, for the three-term solution,

$$A_1 = -5.12 \times 10^{-3}q_o/D; \; A_2 = 5.86 \times 10^{-5}q_o/D;$$
$$A_3 = -2.19 \times 10^{-6}q_o/D; \; B_1 = 1.75 \times 10^{-3}q_o/D;$$
$$B_2 = -1.31 \times 10^{-5}q_o/D; \; B_3 = 3.62 \times 10^{-7}q_o/D;$$
$$C_1 = 8.28 \times 10^{-4}q_o/D; \; C_2 = -1.36 \times 10^{-5}q_o/D;$$
$$C_3 = 5.69 \times 10^{-7}q_o/D; \; D_1 = -4.84 \times 10^{-4}q_o/D;$$
$$D_2 = 4.30 \times 10^{-6}q_o/D; \; D_3 = -1.21 \times 10^{-7}q_o/D.$$

The corresponding central deflection turns out to be 0.475 mm.

Problem 21

(a)   Due to symmetry about the x-axis, C_m and D_m are zero. Further, only odd-m terms occur in the series for the deflection, due to symmetry about $x = a/2$. The constants A_m and B_m are given by

$$A_m = -B_m p_m \tanh p_m; \quad B_m = -\frac{2Ma^2}{m^3\pi^3 D \cosh p_m}$$

(b)   Here A_m and B_m are zero. Again only odd-m terms are present. C_m and D_m turn
out to be

$$C_m = -\frac{2Ma^2 p_m}{m^3 \pi^3 D \sinh p_m \tanh p_m}$$
$$D_m = -C_m \tanh p_m / p_m$$

Problem 22

The given moment loads can be obtained by superposing (a) $M_1(x) = M_2(x) = 2\,\text{Nm/m}$
and (b) $M_1(x) = M_2(x) = -1\,\text{Nm/m}$.

Due to (b), the central deflection will be zero. Thus, considering the
series corresponding to (a) above, and taking terms up to $m = 5$, one gets
$w_{\text{central}} = 0.122$ mm.

Appendix
Solutions for Problems–Chapter 5

Problem 2

This problem can be solved easily by starting from the general equation for the strain energy as follows.

$$U_e = \frac{1}{2} \iint (\sigma_r \varepsilon_r + \sigma_\theta \varepsilon_\theta) \, 2\pi r \, dr \, dz$$

which on simplification, yields

$$U_e = \pi D \int \left(r w_{,rr}^2 + 2\mu w_{,rr} w_{,r} + \frac{w_{,r}^2}{r} \right) dr$$

Problem 3

(a) This is similar to Example 5.1. C_3 and C_4 are zero while C_1 and C_2 are obtained from the clamped edge conditions. One finally obtains

$$w = \frac{q_o(a^2 - r^2)^2}{64D}.$$

(b) Superposition of the solution for a simply supported plate under uniform load (Example 5.1) and that for outer-edge moments (Example 5.4) with enforcement of the zero slope condition at the outer edge leads to $M_o = -qa^2/8$ and the deflection function as in (a) above.

© The Author(s) 2021
K. Bhaskar and T. K. Varadan, *Plates*,
https://doi.org/10.1007/978-3-030-69424-1

Problem 4

Successive integration of Eq. (5.14) with $q = q_o r/a$ leads to

$$w = \frac{q_o r^5}{225aD} + C_1 + C_2 r^2 + C_3 \ln \frac{r}{a} + C_4 r^2 \ln \frac{r}{a}$$

where

$$C_1 = \frac{(6+\mu)}{150(1+\mu)} \frac{q_o a^4}{D}; \qquad C_2 = -\frac{(4+\mu)}{90(1+\mu)} \frac{q_o a^2}{D};$$

$$C_3 = C_4 = 0$$

Problem 5

Here

$$q = q_o(1 - r/a)$$

which is a uniform load q_o (as in Example 5.2) minus a load varying linearly from zero at the centre to q_o at the outer periphery (as in Problem 4 above). The final solution is thus easily obtained by superposition of the solutions for the above two cases.

Problem 6

The general solution of Example 5.1 holds good for these cases. Enforcement of the appropriate boundary conditions yields

For (a):

$$C_1 = 7.067 \times 10^{-4}; C_2 = -3.627 \times 10^{-3};$$
$$C_3 = -2.757 \times 10^{-4}; C_4 = -5.759 \times 10^{-4};$$

and hence $w_{\text{max}} = 0.974$ mm at the inner periphery.

For (b):

$$C_1 = 6.000 \times 10^{-4}; C_2 = -3.201 \times 10^{-3};$$
$$C_3 = 1.119 \times 10^{-4}; C_4 = -5.759 \times 10^{-4};$$

and hence $w_{\text{max}} = 0.402$ mm at the inner periphery.

Problem 7

Proceeding as in Example 5.3, and applying the clamped edge conditions, one finally obtains.

$$w = \frac{P}{16\pi D}\left(2r^2 \ln \frac{r}{a} + a^2 - r^2\right)$$

Problem 8

The given load can be expressed as.

$$q = \frac{5b - 3a}{b - a} - \frac{2r}{b - a}$$

corresponding to which the deflection is obtained as

$$w = \left(\frac{5b - 3a}{b - a}\right)\frac{r^4}{64D} - \frac{2r^5}{225D(b - a)} + C_1 + C_2 r^2 + C_3 \ln \frac{r}{a} + C_4 r^2 \ln \frac{r}{a}$$

The constants are obtained using the boundary conditions as

$$C_1 = -7.56/D; \quad C_2 = 48.02/D; \quad C_3 = -5.15/D; \quad C_4 = -73.38/D$$

The maximum bending moment is M_r at the inner periphery which has a value of -120.94 Nm/m. Using this to calculate the maximum bending stress and equating it to the allowable value, one gets the thickness required to be 3.11 mm.

Problem 9

The free bodies corresponding to the inner loaded area and the outer ring are as shown. The shear force at $r = 0.1$ m is easily calculated as

$$Q_r = -\frac{1000 \times \pi \times 0.1^2}{2\pi \times 0.1} = -50 \, \text{N/m}.$$

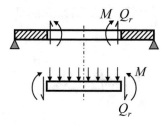

The solution for the deflection of the inner loaded area is obtained using a superposition of the solution for a simply supported circular plate under uniform load (Example 5.2) with that for a plate under axisymmetric moments alone (Example 5.4). This gives the deflection of the inner circular area (*with respect to its outer boundary*) as

$$w_{inner} = \frac{q_o(0.1^2 - r^2)}{64D}\left[\left(\frac{5+\mu}{1+\mu}\right)0.1^2 - r^2\right] + \frac{M}{2D(1+\mu)}(0.1^2 - r^2)$$

For the outer unloaded portion, the solution is exactly as in Example 5.6, except that here Q_r at the inner boundary is –50 N/m. The constants C_1 to C_4 are obtained as

$$C_1 = -6.64 \times 10^{-4} - 1 \times 10^{-5}M;$$
$$C_2 = -4.15 \times 10^{-3} + 6.25 \times 10^{-5}M$$
$$C_3 = -1.67 \times 10^{-4} + 3.71 \times 10^{-5}M;$$
$$C_4 = 3.05 \times 10^{-3}$$

The unknown moment M has to be found out by enforcing the continuity of slope for the two free bodies at $r = 0.1$ m. This yields $M = 4.92$ N m/m.

The maximum deflection and the maximum stress are then obtained as 0.575 mm and 2.62 MPa, respectively, at the centre of the plate.

Problem 10
The central deflection for the present case is the same as that due to an axisymmetric ring load of a total magnitude of 80 N. Using the solution for the maximum deflection as in Example 5.6 with $P = 80/(2\pi \times 0.6)$ N/m, and taking $\mu = 0.3$, one gets the flexural rigidity required for a limiting deflection of 4 mm to be 500.07 Nm.

Problem 11
A look at the free bodies as shown will enable one to understand the methodology of the solution.

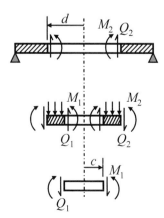

It is easy to see that Q_1 is zero, while Q_2 can be calculated as

$$Q_2 = -\frac{q_o \pi (d^2 - c^2)}{2\pi d}$$

The solutions for the three free bodies can be easily obtained in terms of M_1 and M_2, which are then found out by enforcing the continuity of slope at $r = c$ and $r = d$.

Appendix
Solutions for Problems–Chapter 6

Problem 1

(a) A 75% decrease. (b) A 100% increase. (c) No change, since both Young's modulus and mass density of steel are approximately thrice their values for aluminium.

Problem 3

Proceeding as in Example 6.2, one gets two frequency equations corresponding to y-symmetric modes and y-antisymmetric modes, respectively, as

$$(\alpha^2 + \beta^2) \cosh \frac{\alpha b}{2} \cos \frac{\beta b}{2} = 0$$

$$\text{with} \quad B_m = D_m = 0, \quad A_m/C_m = -\cos \frac{\beta b}{2} \Big/ \cosh \frac{\alpha b}{2}$$

and

$$(\alpha^2 + \beta^2) \sinh \frac{\alpha b}{2} \sin \frac{\beta b}{2} = 0$$

$$\text{with} \quad A_m = C_m = 0, \quad B_m/D_m = -\sin \frac{\beta b}{2} \Big/ \sinh \frac{\alpha b}{2}$$

These can be simplified as

$$\cos \frac{\beta b}{2} = 0 \quad \text{with} \quad \sin \frac{m \pi x}{a} \cos \beta y \text{ as the mode shape}$$

and

$$\sin \frac{\beta b}{2} = 0 \quad \text{with} \quad \sin \frac{m \pi x}{a} \sin \beta y \text{ as the mode shape.}$$

© The Author(s) 2021
K. Bhaskar and T. K. Varadan, *Plates*,
https://doi.org/10.1007/978-3-030-69424-1

The frequency equation obtained by Navier's method (Example 6.1) gives the first three frequencies as

$$\frac{\bar{\rho}\omega_{mn}^2}{D} = \frac{25\pi^4}{16b^4}, \frac{4\pi^4}{b^4}, \frac{169\pi^4}{16b^4}$$

corresponding to $m = n = 1$; $m = 2, n = 1$; and $m = 3, n = 1$, respectively. All these are y-symmetric modes and can be shown to satisfy the first frequency equation as given by Levy's method.

The mode shapes yielded by Navier's and Levy's methods can also be shown to be the same by means of a simple coordinate transformation corresponding to the change of origin.

Problem 4
The frequency equation is obtained as

$$\begin{vmatrix} \cosh\frac{\alpha a}{2} & -\sinh\frac{\alpha a}{2} & \cos\frac{\beta a}{2} & -\sin\frac{\beta a}{2} \\ \alpha^2\cosh\frac{\alpha a}{2} & -\alpha^2\sinh\frac{\alpha a}{2} & -\beta^2\cos\frac{\beta a}{2} & \beta^2\sin\frac{\beta a}{2} \\ \cosh\frac{\alpha a}{2} & \sinh\frac{\alpha a}{2} & \cos\frac{\beta a}{2} & \sin\frac{\beta a}{2} \\ \alpha\sinh\frac{\alpha a}{2} & \alpha\cosh\frac{\alpha a}{2} & -\beta\sin\frac{\beta a}{2} & \beta\cos\frac{\beta a}{2} \end{vmatrix} = 0$$

It can be verified that the above determinant vanishes (actually becomes a small value compared to any of its elements) for the given value of the frequency parameter, for $m = 1$.

The mode shape is given by

$$w(x, y) = \left(\cosh\frac{5.79y}{a} + 1.01\sinh\frac{5.79y}{a} \right.$$
$$\left. + 32.24\cos\frac{3.71y}{a} - 9.45\sin\frac{3.71y}{a} \right) \sin\frac{\pi x}{a}$$

This varies in the y-direction as shown below.

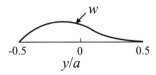

Appendix
Solutions for Problems–Chapter 7

Problem 1

There will be two nodal lines dividing the plate into three squares of side b.

The change in the buckling load is

(a)   20% increase;
(b)   23.5% increase
(c)   33.1% increase;
(d)   No change.

Problem 2

The lowest buckling load is obtained as

$$N_{x\ cr} = -k\frac{\pi^2 D}{b^2}$$

where k is 4.5, 4.17 and 4.08 for $a/b = \sqrt{2}, \sqrt{6}$ and $\sqrt{12}$, respectively, and approaches 4 as one considers further transition points. Thus, for $a/b > \sqrt{12}$, one can take k to be 4 with very little error.

© The Author(s) 2021
K. Bhaskar and T. K. Varadan, *Plates*,
https://doi.org/10.1007/978-3-030-69424-1

Appendix
Solutions for Problems–Chapter 8

Problem 1 Let the plate occupy the domain $0 \leq x \leq a, 0 \leq y \leq b$. A set of admissible functions for the deflection, satisfying the essential boundary conditions alone, are:

a. Simply supported at $x = 0, a$, clamped at $y = 0, b$.

(i) $x(x-a)y^2(y-b)^2$

(ii) $x(x-a)(x-a/2)y^2(y-b)^2(y-b/2)$

(iii) $x(x-a)(x-a/3)(x-2a/3)y^2(y-b)^2(y-b/3)(y-2b/3)$

b. Simply supported at $x = a$, free at $y = b$, clamped at the other two edges.

(i) $x^2(x-a)y^2$

(ii) $x^2(x-a)(x-a/2)y^2(y-2b/3)$

(iii) $x^2(x-a)(x-a/3)(x-2a/3)y^2(y-b/3)(y-2b/3)$

c.   Clamped at $x = 0$, free at $y = b$, simply supported at the other two edges.

(i) $x^2(x-a)y$

(ii) $x^2(x-a)(x-a/2)y(y-2b/3)$

(iii) $x^2(x-a)(x-a/3)(x-2a/3)y(y-b/3)(y-2b/3)$

Problem 2
In terms of trigonometric functions, one gets

a.   (i) $\sin(\pi x/a)\sin^2(\pi y/b)$
     (ii) $\sin(2\pi x/a)\sin(\pi y/b)\sin(2\pi y/b)$
     (iii) $\sin(3\pi x/a)\sin(\pi y/b)\sin(3\pi y/b)$
b.   (i) $\sin(\pi x/a)\sin(\pi x/2a)\sin^2(\pi y/3b)$
     (ii) $\sin(2\pi x/a)\sin(\pi x/2a)\sin(4\pi y/3b)\sin(\pi y/3b)$
     (iii) $\sin(3\pi x/a)\sin(\pi x/2a)\sin(7\pi y/3b)\sin(\pi y/3b)$
c.   (i) $\sin(\pi x/a)\sin(\pi x/2a)\sin(\pi y/3b)$
     (ii) $\sin(2\pi x/a)\sin(\pi x/2a)\sin(4\pi y/3b)$
     (iii) $\sin(3\pi x/a)\sin(\pi x/2a)\sin(7\pi y/3b)$

© The Author(s) 2021
K. Bhaskar and T. K. Varadan, *Plates*,
https://doi.org/10.1007/978-3-030-69424-1

Problem 3

The total potential energy per unit length of the strip is obtained as

$$\Pi = \frac{D}{2} \int_0^a w_{,xx}^2 \, dx - P \, w|_{x=a/2}$$

Using the admissible function $w = A \sin \frac{\pi x}{a}$, one obtains the maximum deflection to be $0.02053 Pa^3/D$. The exact value is $Pa^3/48D$ or $0.02083 Pa^3/D$.

Problem 4

Taking a plate bounded by $x = 0$, a and $y = 0$, b, one gets

$$\int_{y=0}^b \int_{x=0}^a w_{,xx} w_{,yy} \, dx \, dy = \int_{y=0}^b \underbrace{w_{,yy} w_{,x}\Big|_{x=0}^a}_{\text{Equal to zero}} dy - \int_{x=0}^a \int_{y=0}^b w_{,x} w_{,xyy} \, dx \, dy$$

$$= - \int_{x=0}^a \underbrace{w_{,x} w_{,xy}\Big|_{y=0}^b}_{\text{Equal to zero}} dx + \int_{x=0}^a \int_{y=0}^b w_{,xy}^2 \, dy \, dx$$

and hence,

$$\iint (w_{,xx} w_{,yy} - w_{,xy}^2) dx \, dy = 0.$$

Problem 6

(a) Taking the origin to be at a corner of the plate and using $w = A \sin \frac{\pi x}{a} \sin \frac{\pi y}{a}$, one gets

$$U_e = \frac{\pi^4 D A^2}{2a^2}; \qquad V = -2M \int_0^a w_{,y}\Big|_{y=0} dx = -4MA$$

From these, one can find out A using the principle of minimum potential energy. The maximum slope is obtained as $(4Ma/\pi^3 D)$ occurring at $(a/2, 0)$.

(b) Using $w = A \sin \frac{\pi x}{a} \sin \frac{2\pi y}{a}$, one gets

$$U_e = \frac{25\pi^4 D A^2}{8a^2}; \qquad V = -8MA$$

and the maximum slope to be of magnitude $(64Ma/25\pi^3 D)$, which occurs at the end-points as well as the mid-point of the line $x=a/2$.

Problem 7

Let us consider the x-component of M first. Corresponding to this alone, the deflected shape will be antisymmetric about the x-axis and symmetric about the y-axis.

A suitable admissible function can be taken as

$$w = A(x^2 - a^2)^2 y(y^2 - a^2)^2$$

The strain energy U_e can be calculated (using the result of Problem 4 to great advantage) to be $10.36DA^2a^{16}$.

The potential energy of the moment is

$$V = -\frac{M}{\sqrt{2}} w_{,y}\Big|_{(0,0)} = -\frac{MAa^8}{\sqrt{2}}.$$

Rayleigh–Ritz method yields $A = 0.03414\, M/Da^8$. The deflection at $(a/2, -a/2)$ turns out to be $0.0054\, Ma/D$ upward due to the x-component alone and is twice this value for the given total moment M.

Problem 8

Taking the origin of the coordinate system at the top-left corner of the plate, and using the admissible function

$$w = A \sin \frac{2\pi x}{a} \sin \frac{2\pi y}{b},$$

one gets

$$U_e = \frac{2\pi^4 D(a^2 + b^2)^2 A^2}{a^3 b^3}; \quad V = 4PA$$

Hence, the deflection under any of the loads is of magnitude $Pa^3 b^3/\pi^4 D(a^2 + b^2)^2$.

Problem 10

(a) With the origin at a corner of the plate, the chosen admissible function is

$$w = A \sin \frac{\pi x}{a} \sin \frac{\pi y}{a}$$

which satisfies all the boundary conditions at the outer boundary. Those at the inner boundary are natural boundary conditions and are not satisfied.

Considering only a quadrant of the plate, and carrying out all integrations over suitable sub-domains, one gets

$$\Pi = \frac{A^2}{576a^2}[2\pi^2 D(32\pi^2 - 27 - 12\sqrt{3}\pi\mu) - P_{cr}a^2(32\pi^2 + 27)]$$

where N_x and N_y are taken as $-P_{cr}$.

Hence,

$$P_{cr} = \left(\frac{2\pi^2 D}{a^2}\right)\left(\frac{32\pi^2 - 12\sqrt{3}\pi\mu - 27}{32\pi^2 + 27}\right)$$

while that for a solid plate is simply $(2\pi^2 D/a^2)$. The second bracketed quantity in the above expression is always less than 1; it has a value of 0.785 for $\mu = 0.3$.

(b) Continuing with the same admissible function as above with A being a function of time as well, and carrying out calculations for one quadrant again, one gets

$$U = \frac{A^2}{576a^2}[2\pi^2 D(32\pi^2 - 27 - 12\sqrt{3}\pi\mu)]$$

$$T = \frac{\bar{\rho}a^2 \dot{A}^2}{1152\pi^2}(32\pi^2 - 27 - 12\sqrt{3}\pi)$$

Hence,

$$\omega^2 = \left(\frac{4\pi^4 D}{\bar{\rho}a^4}\right)\left(\frac{32\pi^2 - 12\sqrt{3}\pi\mu - 27}{32\pi^2 - 12\sqrt{3}\pi - 27}\right)$$

The first bracketed quantity of the above equation is the value for a solid plate. The second bracketed quantity reflects the influence of the hole and is always greater than 1; for $\mu = 0.3$, it is 1.204.

Problem 11
The energies are obtained as

$$U_e = \frac{\pi^3 D}{48a^2}[108\pi A^2 - 112(7 + 3\mu)AB + 675\pi B^2]$$
$$T = \frac{\bar{\rho}a^2}{48\pi}[9\pi \dot{A}^2 - 16\dot{A}\dot{B} + 9\pi \dot{B}^2]$$

Proceeding further, one gets the following frequency equation (for $\mu = 0.3$):

$$2941.75\lambda^4 - 249,891\lambda^2 + 2.9051 \times 10^6 = 0$$

where $\lambda^2 = \frac{\bar{\rho}a^4\omega^2}{\pi^4 D}$.

The corresponding frequencies and mode shapes are given by

$$\lambda_1^2 = 9.431 \quad \text{with} \quad A/B = 5.052$$
$$\lambda_2^2 = 75.52 \quad \text{with} \quad A/B = 0.090$$

(For a plate of uniform thickness h, the two values of λ^2 are 4 and 25.)

Problem 12
With the origin at a corner of the plate, a suitable admissible function is

$$w = A(t)x^2(x - a)^2 y^2(y - a)^2.$$

This gives

$$U_e = \frac{2Da^{14}A^2}{1225}; \quad T = \frac{\overline{\rho}a^{18}\dot{A}^2}{793,800} \quad \text{and} \quad \omega^2 = \frac{1296D}{\overline{\rho}a^4}.$$

The mass lumped at the centre increases the kinetic energy by

$$\frac{1}{2}\frac{\overline{\rho}a^2}{3}\dot{w}^2_{\text{centre}}$$

i.e. $\dfrac{\overline{\rho}a^{18}\dot{A}^2}{393216}$

while the strain energy remains unaffected. The new frequency is given by $\omega^2 = \frac{429.32D}{\overline{\rho}a^4}$.

Problem 13
A suitable admissible function satisfying all the boundary conditions (for a plate with the origin at a corner) is

$$w = A \sin \frac{\pi x}{a} \sin \frac{\pi y}{a}.$$

Corresponding to this, one gets

$$U_e = \frac{\pi^4 D A^2}{2a^2}; \quad V = -\int\limits_{y=a/3}^{2a/3} P\, w|_{x=a/2} dy = -\frac{P A a}{\pi}$$

and hence, by Rayleigh–Ritz method, $w_{\text{max}} = \frac{Pa^3}{\pi^5 D}$.

To use Galerkin's method, one has to look at the line load as the limiting case of a distributed patch load of width, say u.

The residual is given by

$$R = (D\nabla^4 w - P/u) \quad \text{within the patch area}$$
$$= D\nabla^4 w \quad \text{else where}$$

with due substitution for w in terms of the admissible function.

Proceeding further, and rearranging the terms for a quadrant of the plate, one gets

$$
\int_0^{\frac{a}{2}} \int_0^{\frac{a}{2}} D\nabla^4 w \sin\frac{\pi x}{a} \sin\frac{\pi y}{a} dx\,dy - \frac{P}{u} \int_{x=\frac{a-u}{2}}^{\frac{a}{2}} \int_{y=\frac{a}{3}}^{\frac{a}{2}} \sin\frac{\pi x}{a} \sin\frac{\pi y}{a} dx\,dy = 0
$$

Evaluating the above integrals and taking limits as $u \to 0$, one gets the same answer as by Rayleigh–Ritz method above.

Problem 14

This is a straightforward application of Galerkin's method to the governing equation with $N_x = -ph$ and $N_y = -P_{cr}$. The residual is given by

$$
R = D\nabla^4 w + phw_{,xx} + P_{cr}w_{,yy}
$$

with due substitution for w.
    Proceeding further, one gets

$$
P_{cr} = \frac{32\pi^2 D}{3a^2} - ph.
$$

The above answer has to be interpreted as follows:

(a) When $ph < 32\pi^2 D/3a^2$, a compressive force of magnitude as given above causes buckling.
(b) When $ph = 32\pi^2 D/3a^2$, the plate is already in a state of neutral equilibrium and cannot take any compressive force in the y-direction.
(c) A value of $ph > 32\pi^2 D/3a^2$ is meaningful only if the force along the y-direction is tensile.

Problem 15

(a) The given function satisfies all the boundary conditions and is suitable for use with Galerkin's method.
    Proceeding further, one gets

$$
\int_0^a (Dw_{,xxxx} - N_{x\,cr}w_{,xx})(a^3 x - 3ax^3 + 2x^4)dx = 0
$$

which, after due substitution for w, yields N_{xcr} to be $-21D/a^2$. The exact solution, by analogy with a corresponding column, is $-20.19D/a^2$.

(b) It can be verified that the area of the tapered cross section is the same as that of the strip of uniform thickness h_o. The flexural rigidity is now a function of x and is given by

$$D(x) = \frac{8x^3 D_o}{c^3} \quad \text{for } x \leq c$$

$$= \frac{8(a-x)^3 D_o}{(a-c)^3} \quad \text{for } x \geq c$$

where $D_o = \frac{Eh_o^3}{12(1-\mu^2)}$.

Proceeding as in (a), and carrying out the integrations in the two regions separately, one gets

$$N_{x \, cr} = -\frac{8D_o(5 + 15s - 12s^2 - 20s^3 + 15s^4)}{a^2} \quad \text{where } s = c/a.$$

For maximum N_{xcr}, one has to take the first derivative of the RHS above with respect to s and equate it to zero. This yields three values of s, out of which only 0.392 is meaningful—the other two values fall outside the interval $(0, 1)$. The value of 0.392 corresponds to a maximum value of the buckling load as can be easily verified by means of a plot. This maximum buckling load is obtained as $-65.48D_o/a^2$, which is more than three times that for a uniformly thick strip of the same weight. For $c=a/2$, one gets $-63.5D_o/a^2$.

Problem 16
Starting with a general approximation for the deflection as

$$w = A(t)(x^2 + C_1 x^3 + C_2 x^4)$$

which satisfies the clamped edge conditions at the end $x = 0$, and enforcing the free edge conditions at the other end $x = a$, one can find out the constants. The resulting admissible function is given by

$$w = A(t)\left(x^2 - \frac{2x^3}{3a} + \frac{x^4}{6a^2}\right).$$

The residual is given by

$$R = D w_{,xxxx} + \bar{\rho} \, \ddot{w}$$

with due substitution for w. Galerkin's methodology yields

$$\omega^2 = \frac{12.46D}{\bar{\rho}a^4} \cdot \text{(The exact solution is } \omega^2 = \frac{12.36D}{\bar{\rho}a^4}.\text{)}$$

Problem 17

The conditions to be satisfied are:
$$w = M_r = 0$$
at $r = a$:
$$\text{i.e. } w = w_{,rr} + \mu w_{,r}/r = 0$$
at $r = 0$: $w_{,r} = 0$.

While those at the outer periphery are boundary conditions, that at the centre of the plate is a condition required to satisfy internal compatibility of deformation.

Starting with a general polynomial approximation for the deflection as

$$w = C_1 + C_2 r + C_3 r^2 + r^3$$

and finding the constants by means of the conditions above, one gets the admissible function as

$$w = A(2.6r^3 - 6.9ar^2 + 4.3a^3) \text{ for } \mu = 0.3.$$

The residual is given by

$$R = \frac{D}{r}\frac{d}{dr}\left[r\frac{d}{dr}\left\{\frac{1}{r}\frac{d}{dr}\left(r\frac{dw}{dr}\right)\right\}\right] - q$$

where $q = 1000$ N/m^2 for $r \leq 0.1$ m and zero elsewhere.

The Galerkin integral has to be evaluated separately in the two regions. The final answer is $A = 0.002015$ m^{-2}. The maximum deflection and the maximum bending stress turn out to be 0.5545 mm (as against the exact value of 0.5745 mm) and 2.224 MPa (exact value = 2.617 MPa), respectively.

Index

© The Author(s) 2021
K. Bhaskar and T. K. Varadan, *Plates*,
https://doi.org/10.1007/978-3-030-69424-1

Printed in the United States
by Baker & Taylor Publisher Services